Sustainable Networks in Smart Grid

Sustainable Networks in Smart Grid

Edited by

B.D. Deebak
School of Computer Science and Engineering,
Vellore Institute of Technology, Vellore, India

Fadi Al-Turjman
Department of Artificial Intelligence Engineering,
Research Center for AI and IoT, AI and Robotics Institute,
Near East University, Mersin, Turkey

ELSEVIER

ACADEMIC PRESS
An imprint of Elsevier

Academic Press is an imprint of Elsevier
125 London Wall, London EC2Y 5AS, United Kingdom
525 B Street, Suite 1650, San Diego, CA 92101, United States
50 Hampshire Street, 5th Floor, Cambridge, MA 02139, United States
The Boulevard, Langford Lane, Kidlington, Oxford OX5 1GB, United Kingdom

ISBN: 978-0-323-85626-3

For Information on all Academic Press publications
visit our website at https://www.elsevier.com/books-and-journals

Publisher: Charlotte Cockle
Acquisitions Editor: Lisa Reading
Editorial Project Manager: Michelle Fisher
Production Project Manager: Kamesh Ramajogi
Cover Designer: Vicky Pearson Esser

Typeset by MPS Limited, Chennai, India

Contents

List of contributors

Chadi Altrjman Department of Chemical Engineering, Waterloo University, Waterloo, Canada

Fadi Al-Turjman Department of Artificial Intelligence Engineering, Research Center for AI and IoT, AI and Robotics Institute, Near East University, Mersin, Turkey

Sinem Alturjman Department of Artificial Intelligence Engineering, Research Center for AI and IoT, AI and Robotics Institute, Near East University, Mersin, Turkey; Department of Artificial Intelligence Engineering, Near East University, Mersin, Turkey

T. Anitha Department of Computer Science and Engineering, Anil Neerukonda Institute of Technology and Sciences, Visakhapatnam, India

A. Anupama Department of Computer Science and Engineering, Anil Neerukonda Institute of Technology and Sciences, Visakhapatnam, India

Barakat A. Dawood Computer Engineering Department, Near East University, Mersin, Turkey; Department of Artificial Intelligence Engineering, Research Center for AI and IoT, AI and Robotics Institute, Near East University, Mersin, Turkey

B.D. Deebak School of Computer Science and Engineering, Vellore Institute of Technology, Vellore, India

Adedoyin A. Hussain Computer Engineering Department, Near East University, Mersin, Turkey; Department of Artificial Intelligence Engineering, Research Center for AI and IoT, AI and Robotics Institute, Near East University, Mersin, Turkey

B. Raja Koti Department of CSE, GITAM Institute of Technology, Vishakhapatnam, India

D. Siva Krishna Department of CSE, GMR Institute of Technology, Srikakulam, India

P. Muralidhara Rao School of Computer Science and Engineering, Vellore Institute of Technology, Vellore, India

B. Padma Gayatri Vidya Parishad College for Degree and PG Courses (A), Visakhapatnam, India

B. Raj Koti GITAM Institute of Technology, GITAM Deemed to be University, Visakhapatnam, India

G.V.S. Raj Kumar GITAM Institute of Technology, GITAM Deemed to be University, Visakhapatnam, India

T. Ramathulasi Department of Computer Science and Engineering, Mother Theresa Institute of Engineering & Technology, Palamaner, Chittoor, India

A.V.H. Sai Prasad GITAM Institute of Technology, GITAM Deemed to be University, Visakhapatnam, India

P. Saraswathi Department of Computer Science and Engineering, GITAM Institute of Technology, Visakhapatnam, India

Reddi Sivaranjani Department of Computer Science and Engineering, Anil Neerukonda Institute of Technology and Sciences, Visakhapatnam, India

Y. Vishnu Srinivasa Murthy School of Computer Science and Engineering, Vellore Institute of Technology, Vellore, India

Hadi Zahmatkesh Department of Mechanical, Electronic and Chemical Engineering, OsloMet—Oslo Metropolitan University, Oslo, Norway

Chapter 1

Smart Grid communication and information technologies for cyber security, data privacy, and policy issues

G.V.S. Raj Kumar[1], A.V.H. Sai Prasad[1], B. Padma[2] and B. Raj Koti[1]
[1]*GITAM Institute of Technology, GITAM Deemed to be University, Visakhapatnam, India,*
[2]*Gayatri Vidya Parishad College for Degree and PG Courses (A), Visakhapatnam, India*

1.1 Introduction to Smart Grids

A Smart Grid is famed as a "Power grid" that utilizes data and communication automation to collect and work on data like data about the practices of suppliers and consumers within an automatic vogue to enhance the fiscal and defensible of the assembly and dispensation of power.

1.1.1 What is a Smart Grid?

An intelligent grid eases the switch from traditional energy to renewable energy. The grid permits easy access to integrate it into the grid by obtaining a renewable energy source in the environment. The Smart Grid aids higher performance of highly changeable renewable sources of energy, such as wind generation power and solar power energy. The detailed interpretation of communication construction in SG as depicted in Fig. 1.1.

1.1.2 Motivations/objectives of the Smart Grids

1.1.2.1 The rationale for Smart Grid technology

Smart Grid technologies manifest ways to confront these challenges and develop a cleaner energy supply that is more methodical, cheaper, and highly tenable. Fig. 1.2 shows the SG stakeholders involved in electricity system.

Sustainable Networks in Smart Grid. DOI: https://doi.org/10.1016/B978-0-323-85626-3.00008-9

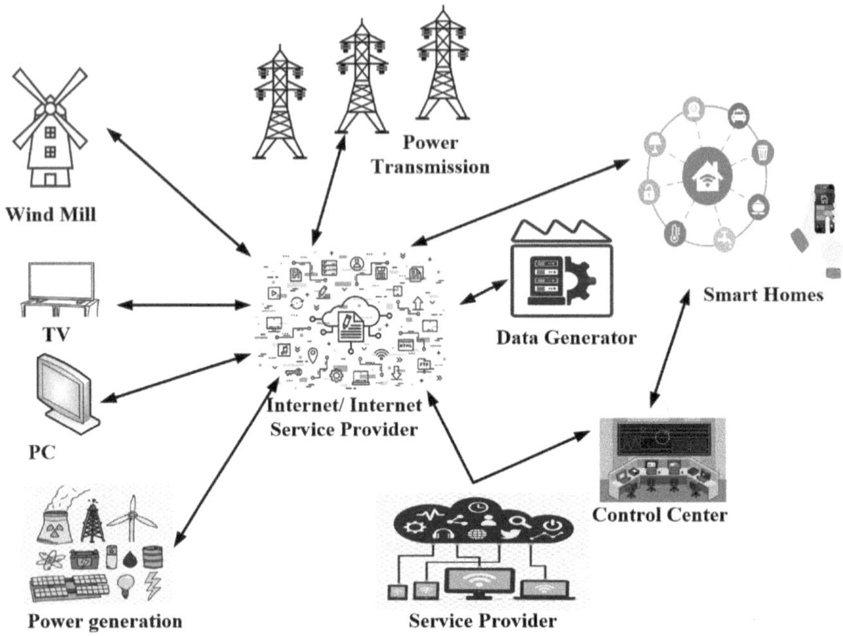

FIGURE 1.1 An interpretation of a communication construction in Smart Grid.

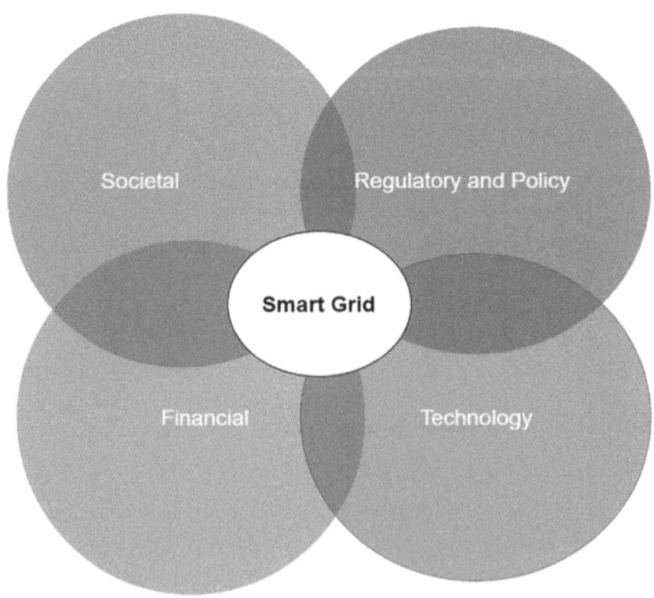

FIGURE 1.2 Smart Grids tie up electricity system stakeholder intentions.

Benefits of the Smart Grid

Coalesced outlying technologies: Smart Grid empower better energy management.

Progressive insistence, supply/insistence retaliation

Stronger power caliber

Minimize carbon emanations.

Pitfalls of Smart Grid

1. Privacy stumbling blocks

    The immense concern is safety with security in a Smart Grid system.

2. Grid fickleness

    Smart Grid network has more detail at its edges, which is related to the arrival point and the end user's meter.

1.1.2.2 Main triggers of the development of Smart Grids

The main four reasons why developing countries require and need Smart Grids are:

Windup power theft: Not such an issue in progressed nations; however, in India, with higher poverty, power stealing is highly practices.

Excessive grade/genuine of power: If homes in progressing nations tie up to the grid, often, the connection is low, and users can only ingress electricity amid a specific period.

Maturation accounts for the cost: The most challenging scenario towards utilities is making the economics of Smart Grids work, but several added customers can assist with the return on investment in China.

Continual energy needs a Smart Grid: Progressive countries were highly fortunate by adding more good power than progressive nations. Utilities will require a Smart Grid to manage bottlenecks caused by intermittency.

1.1.2.3 Power industry enabled by digital technologies

Overlay disruption in the sector, many electric power companies are looking to technological innovations that can sustain their business for the long-drawn-out.

The electric power sector has witnessed an unrivaled change in the former decade, with renovation energy sources being included in the mix. In contrast, digital technologies have alternated the routing energy is generated, supplied, and consumed by businesses and consumers.

A rising number of electric power companies recognize their digital transformation approaches as their precedence. Around the summit, Huawei and manufacturing heads identified the benefits of '5G,

A.I., Big data and cloud computing, those exemplified at the online occasions.

1.1.2.3.1 Putting the smarts into energy

One such specimen of change is the Smart Grid. Instead of manually investigating transmission lines, which is costly and ineffectual, the energy community can use smart technologies to monitor their transmission lines remotely.

1.1.2.3.2 New technologies come into play

For example, the 'State Grid Corporation of China' collaborated with Huawei to develop digital and IoT structures and cloud that reduced the period it inputs to gather and store data.

1.1.2.3.3 Leap now

This sector has relied on old technology for quite some time. That can no longer be the plan of action in the future now, with interference coming to the sector. A change could mean leaping forward; for others, it may deepen their already ongoing transformation efforts.

1.1.2.4 Objectives and motivations addressed across Smart Grids

1.1.2.4.1 Motives and objectives of Smart Grid

The overall aims of Smart Grid are intended to secure a vibrant, continual, and biodegradable system potency that is value and fuel-efficacious.

The aims of originating the Smart Grid are dissimilar from nation to nation for their varying claims and entry points.

Nevertheless, the familiar objectives of a Smart Grid are clear and given below:

Soundness: The Smart Grid will ameliorate durability to interference to bring forth constant and sturdy electricity flow, circumvent expansive flare-up mishaps.

Defended operation: The Smart Grid shall inflate communication networks and the "electricity-grid" data reliability.

Blended system: The Smart Grid shall greatly amalgamate and distribute facts and data of an electricity grid, make use of the sustained rostrum and replica to impart regulated and purified administration.

Accession: The Smart Grid shall revamp holdings, lower outlays and drive precisely.

Bio-energy: The Smart Grid shall work out issues of vitality safety, vitality thrift, carbon dioxide emanation etc.

1.1.3 Smart Grid applications and characteristics

1.1.3.1 Requisitions of a Smart Grid

Deploying digital technology in "smart grids" summons the trust, efficacy, and readiness of the end-users concerning all amenities that count towards the nation's economic tenacious.

The rudimentary oppositions of Smart Grids are:

- They facilitate the adequacy of transmission lines.
- Brisk reclamation in the wake of any unforeseen wreckage in lines-and-feeders.
- Cost depletion.
- Depletion of peak hours.

1.1.3.2 Smart Grid characteristics

The important attributes of Smart Grids are described below:

- Enables informed participation by customers.
- A Smart Grid lodges immense, centralized power-plants and increases the array of customer-centric distributed energy resources.
- Entitles new brands, services, and trade.
- Rightly outlines and manages markets to generate good fortune for consumers to choose between competing services.
- Managers, operators, and consumers entail the pliancy to amend business rules to outfit operating, and market set-up brings forth the power caliber for the range of needs.

1.1.4 Smart Grid infrastructure and challenges

1.1.4.1 Communication network in Smart Grids

This part inspects the communication infrastructure framework in usage in Smart Grids.

Fig. 1.3 manifests an outline of the standard classic architecture of a Smart Grid. Customer- Premises-Network (CPN).

It was owing to the unalike essences of these sites, three discrete networks would-be flourish to outfit: "Home-Area-Networks (HAN)," "Business/Building Area Networks (BAN)," and "Industrial Area Networks (IAN)".

The "HAN" adequately superintends the quest power requirements of the end-customers. This network is contemplated to conjugate elegant electric devices, namely television sets, washing machines, or energy conservation rostrums.

The "BAN," or Commercial Area Network, is a communication framework deliberate to prop up the needs of well-ordered businesses.

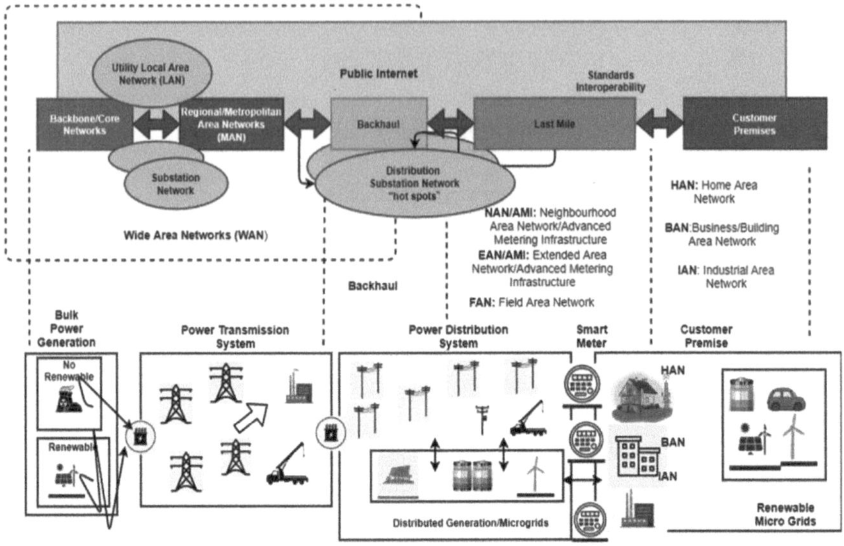

FIGURE 1.3 Complete Smart Grid Communication architecture.

Lastly, the "IAN" can be defined as the communication infrastructure that permits the interdependence and abutment of all machines and appliances are depicted in Fig. 1.4.

1.1.4.1.1 Architectonics and technologies used in CPN

Confide in on the end-user campus network type "HAN", "IAN," or "BAN" various modern technologies may be employed.

1. **Communication networks in the distribution-grid**

   Under the advent of the Smart Grids, dispersive systems have radically shifted. Prior accost by paramount market players "diffusion networks are underneath excessive oppression to meet needs for transmuting regular static grids into contemporary and effectual Smart Grids.
2. **Architectonics and technologies in the distribution-grid**

   Smart Grids have brought many modifications to regular power grids, building the latest technologies and devices to entitle new trademarks that concede finer communication, dispensation and all-inclusive competence.
3. **Communication networks in the transmission-grid**

   Transferal comes about the vast devolve of electrical power from peer group authorities to the dispensation grid via multiple substations, which TSOs usually handle. The prime intent of TSO is to keep going the firmness of the electric grid by weighing up peer group (Endow) and load (Behest) via the transmission network.

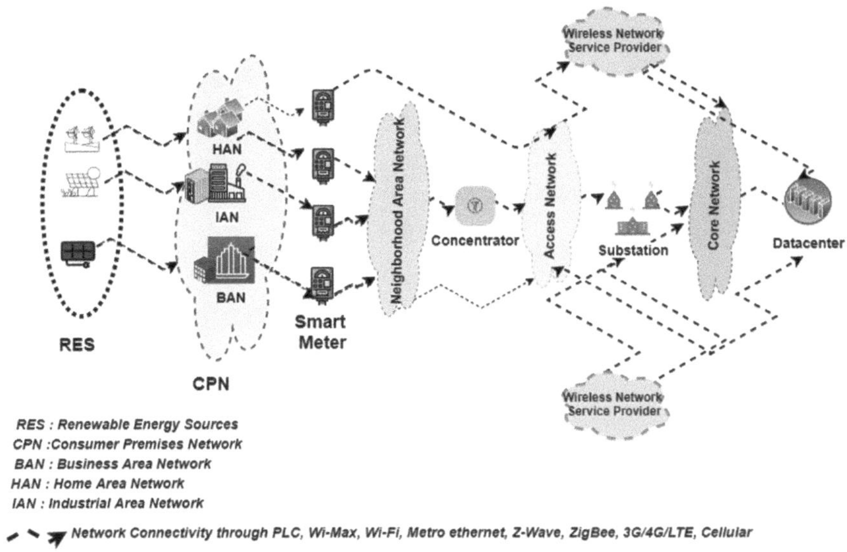

FIGURE 1.4 Basic network architecture.

4. **Communication network unification between generation, transmission and distribution grids**
   Though copious custody and extemporary trade-off protocols are outlined to reciprocate data amid these elements, it is widespread for actors to reciprocate data with everyone using assimilate standards, as it is typically simpler and more authentic.

1.1.4.2 Demand response applications

Demand response set forth a fortune for end-users to take part in a substantial chunk in the electric grid potency, thereby lowering or transferring their energy consumption during peak periods in retribution to time-dependent tariff or other shapes of monetary inducements.

1.1.4.3 Smart Grid intelligent information processing

Towards this, the aim of omnipresent data and crop up Information and Communication Technologies (ICTs) infrastructure is essential towards awareness of "SGI-4.0". In "SGI-4.0", the critical objective of ICTs is to erect an excessive authentic and pliable communication infrastructure. The pivotal transmission and utilities architectonics for Smart Grids near the backdrop of Industry 4.0 is depicted in (Fig. 1.5).

1.1.4.4 Microgrid and integration of energy sources

Electric-Power can bring out at a dispensation magnitude in a microgrid. It typically embraces various tiny power-producing sources and power

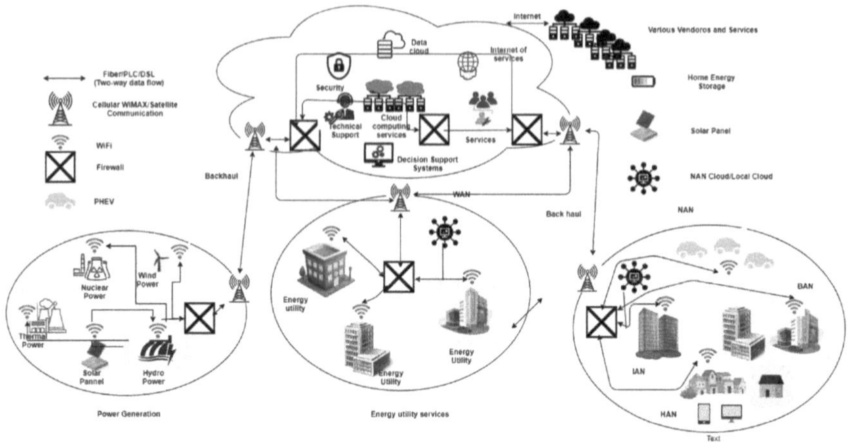

FIGURE 1.5 The pivotal transmission and utilities architectonics for Smart Grids near the backdrop of Industry 4.0.

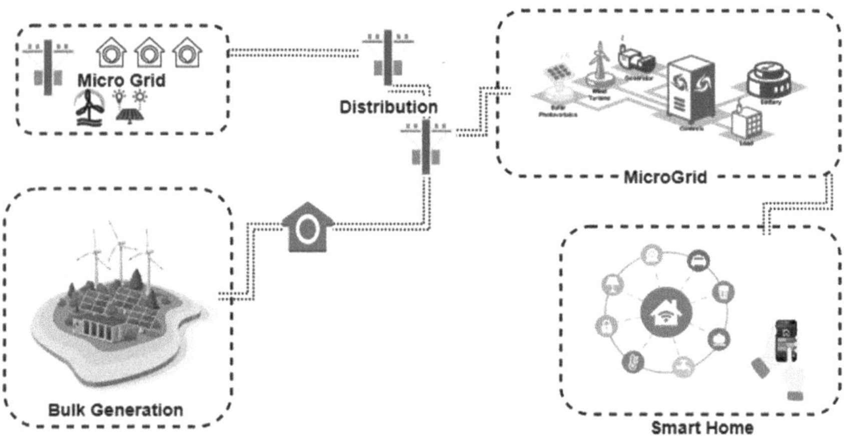

FIGURE 1.6 The theoretical impression of a microgrid.

storehouse systems such-as <Batteries, flywheels, and super capacitors are depicted in Fig. 1.6.

The conventional power supply system is not delineating to subsume power origination and storage at the dispersal level. It also does not disseminate energy sources to impart the customers head-on.

In synopsis, microgrids can be escalating the genuine power supply nearby via agile control of intramural loads and generations. In addition, it can subsume sustainable energy sources that help out to bring down contaminations.

1.1.4.5 Technical challenges in Smart Grids

1.1.4.5.1 Technical challenges

1. **Paucity in grid infrastructure**: The subsist grid network lacks to serve the requirements of eco-clean energy and dispensed generation, which may throw various demands in design, elevation, operation, and perpetuation.
2. **Cyber security**: Connecting the grid to a cyber-network brings about various susceptibilities in the system, and acutely, we are unaware of them. Confess and obliterate such omissions before any security breaking takes place is vital.
3. **Data-management**: SG pervades power networks with vast amounts of meters, sensors-and-controllers. The facts from these segments and other sources such as weather-forecast, safety cameras, etc., inflate the aptness of operators. Figs. 1.7 and 1.8 depicts SPV process behind and way of battery aid.
4. **Communication issues**: We go through a wide range of communication technologies for disposition in SG, but they all possess their impediments. One-off technology has minimal bandwidth while the second works in a finite distance, the third has excessive data loss, and the other has minimal triumph in concealed installations.
5. **Energy management and electric vehicle**: Make use of electric vehicles (EVs) as storage forthcoming is along with the manifesto. Research towards the coherent exertion of electric cars in the course of intervals of peak hour is in progress.

1.2 Evolution of the Power Grid

1.2.1 The Power Grid

An electrical grid is an electrical power system network that consists of the generating plant, transformers, transmission lines, the substation, distribution

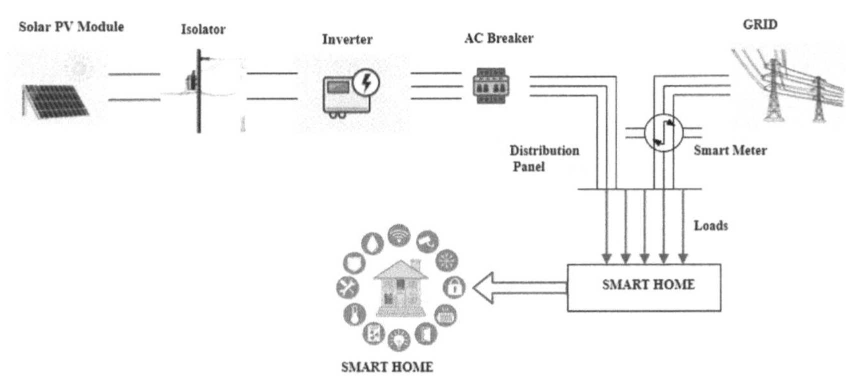

FIGURE 1.7 SPV Process besides no battery aid.

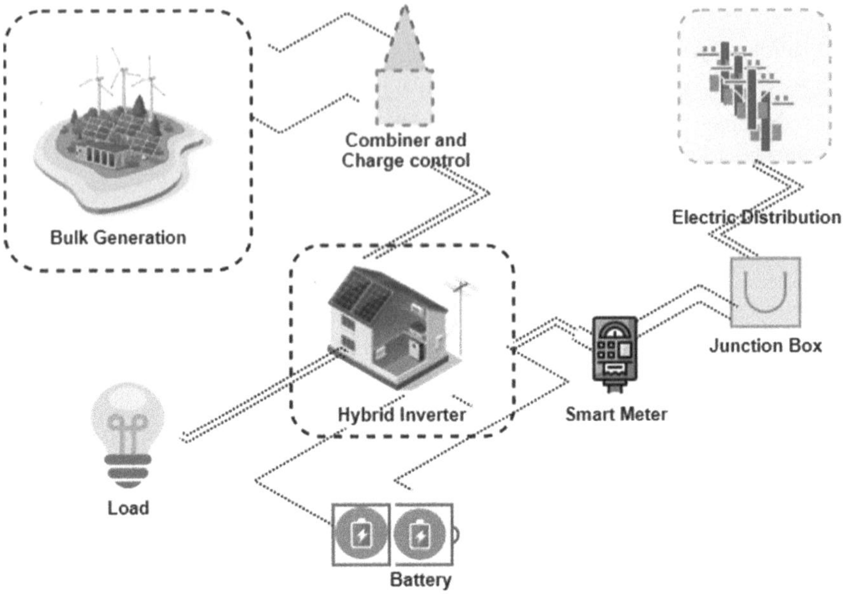

FIGURE 1.8 SPV Process by way of battery aid.

lines and consumers. The Power Grid is a network meant for bringing electricity to users. The Power Grid comprises generator stations, towers, transmission lines, and distinct consumer distribution lines. The generator yields energy and converts energy into a high power for distribution (Amarsingh, Latchman, & Yang, 2014). The electrical grid comprises three major components:

- **Generation**—There are two modes of generation: one is centralized and decentralized. Rooftop solar is an example of decentralized generation. Centralized generation includes natural gas, coal, nuclear, hydro, large solar arrays, and wind farms. The grid is responsible for connecting centralized power to users.
- **Transmission and distribution**—Transmission comprises substations, transformers, and power lines that carry current to the consumption points. When there is a high voltage of electricity, transmission problems are less for long distances. Transmission is realized through powerlines and occurs either underground or overhead.
- **Consumption**—There are different types of consumers, such as residential, manufacturing, and commercial. All these consumers have diverged needs, but electricity delivers vital energy services such as power and light for appliances.

The Power Grid began as an isolated energy generation system in the 1870s. In 1882, Thomas Edison invented the primary central power station in New York City. There exist different types of power plants, and their classification depends on the fuel used. In the initial period of the 20th century, over 4000 individual power utilities were existing, each operating individually. These utilities are functioning with low-voltage "power plants" which served the local customers via short distribution lines. By way of the demand for electric power improves, especially in the post-World War II era, electric efficacies found that it is more important and efficient to connect their transmission systems (Gelazanskas & Gamage, 2014). They might gain the advantages of building jointly-owned and larger generators to serve the combined electricity demands at the lowest prices while evading duplicate energy plants.

The Smart Grid is a natural development of the Power Grid in advanced countries. It is important for making power more accessible and improving the economy of developing countries since the technology enables consistent power quality for commercial industries and factories. Smart Grid was first implemented by the Energy Independence and Security Act of 2007 (EISA-2007) and was approved in Jan 2007 by the U.S. Congress. Substantial industry-wide revolution is also taken place on account of greater interest in distributed energy resources.

The microgrid is the driving change is the third emerging technology. They are completely built out as smaller-scale versions of old grid systems and they are precisely focused on delivery and localized power generation for smaller communities and military bases. Present microgrids are naturally connected to the main power grid. They have enough on-site self-generated power to disconnect from the grid in the context of a key disruption (Alsaidan, Alanazi, Gao, Wu, & Khodaei, 2017).

The fourth industrial uprising, also termed Industry 4.0, has obtained wide discussion and huge consideration from manufacturers and researchers. Industry 4.0 supports smart production procedures and achieves complete automation. It addresses enhancing the manufacturing procedures, enabling the advancement of industrial demand and supply integration. IoT, Cyber-Physical Systems (CPS), Big Data Analytics and Automation remain the main elements of Industry 4.0, a major turning point in the energy field considered the important point of all revolutions. The objective is that energy transmission, generation, and distribution are more organized and reliable with the next-generation hardware and software, which is a consequence of this idea.

1.2.2 Traditional Power Grid and energy management

In the present world of high energy demand, power generation must be increased to fulfill the user's needs and improve their regular life. However, since more consumers exist and the unpredictable nature of the electric load, power demand causes more challenges to the system operator and electric utilities. To solve this issue, system operators have two available choices:

- Augment the size of the network, which requires time to implement and is costly.
- Employ energy management to decrease the opportunity of high demand during peak time.

Energy management is useful for a smarter grid for the following reasons:

- It does not need direct intervention from human beings as it is automated and provides predictions and accurate results.
- It supports the electric utility in optimizing the working of its units and reduces the generation cost.
- It helps the users in managing their load demand and reduces the cost of electricity.
- Increases the load factor and energy efficiency so that the power profile becomes less fluctuating and smoother (Portelinha Júnior, Zambroni de Souza, Castilla, Queiroz Oliveira, & Ribeiro, 2017).
- It conserves resources and reduces pollution and protects the climate.

The main powerful factors to improve current power supply grids are classified into three categories:

1. Increasing reliability, safety, the efficiency of the grid by decreasing demand at peak time.
2. Augmenting flexibility of energy consumption.
3. Permitting consumers to turn either as electrical energy clients while consuming or as electrical power suppliers while producing.

The overall load connected to the power grid can oscillate considerably with time about the first driving factor. For example, when a television program starts, a million watchers switch on their televisions simultaneously, resulting in a rapid upsurge of power consumption. This may result in grid disruption and blackouts. Because of this reason, mathematical and scientific algorithms have been designed to forecast the increase in power consumption, to make correct decisions. Electricity blackouts, power outages and brownouts are very expensive and very painful for users. So, it is beneficial for end-users and energy suppliers to reduce the count and time-span of the high demand periods. To decrease grid maintenance costs, the need for standby generators would be significantly reduced. Also, a reliable and quality service can be offered to the users. Smart grids boost the energy proficiency of the power grid for end-users by scheduling and coordinating the low-priority home devices.

Also, an improved electrical grid reduces CO_2 releases by dropping user energy consumption in peak hours, where electricity generated through power plants emits a huge amount of CO_2 (Uribe-Pérez, Hernández, de la Vega, & Angulo, 2016). The second driving factor minimizes maintenance, operation costs and also optimizes the capital assets. End-users can gain profits from lower-cost generation sources and renewable power sources like wind turbines. Matching I.T. (Information Technology) with renewable energy sources can produce new ways of achieving energy savings and reducing CO_2 releases.

1.2.3 Environmental issues in traditional Power Grids

The traditional power grid was undergoing many issues and challenges, along with a changing generation situation with an increase in remote and renewable generation, an old A.C. transmission set-up, and a worldwide electricity demand (Jerin et al., 2018). When electricity lines on the roads are placed in immature areas, they can distract wetlands, forests, and other green areas. Many pollutants are released into the atmosphere from a coal power plant. These contain Carbon Monoxide (C), Sulfur Dioxide (SO_2), Oxides of Nitrogen (NOx) and finally, Ozone (O_3). Lead, Suspended Particulate Matter, and Non-Methane Hydrocarbons are also emitted. Power plants release mercury, a neurotoxin found in all waterways, and carbon dioxide, which is the most substantial greenhouse gas and causes climate change. These plants also release arsenic, cadmium, beryllium, chromium, cadmium, and nickel. Pre-processing coal can reduce the undesirable compounds in burning gases. The burning of low-sulfur coal reduces SO_2 releases. An electric device, lokes such as a refrigerator, a vacuum cleaner or a computer monitor, is a source of ELF radiation. Electric blankets also expose individuals to extremely low frequency (ELF) radiation. Electrocution on current lines is supposed to be a major danger to bird species around the world. A new study emphasizes avoiding bird electrocution by identifying and correcting high-risk pylons.

The procedures of the current Power Grids are inefficient and limited. Traditional Power Grids lack effective energy storage and backup capacity, and also the supply is also needed to sustain the necessity. So suitable energy and advanced storage technologies are vital for Power Grid. Furthermore, because of the peak load hours, electricity cut-offs, or quality issues concerning readiness and reliability, are issues of concern. Peaker plants run in high demand. However, this technique is ineffective and not economical when the average electricity demand is less than the peak. Also, it is difficult to match the energy supply to the highest demand using these Peaker plants as day-after-day energy demands increase.

A small letdown becomes a fatal effect in Peaker plants to crash the whole grid. Also, traditional Power Grids do not depend on high-quality power. Solar, geothermal, hydro, wind, and biomass are the possible resources for renewable energy. Their integration with Power Grids in the procedure of electric energy generation meaningfully decreases the production of destructive gases into the air to reduce the negative outcomes of environmental change. Solar and wind power are available in various parts of the world, yet, their involvement in the power grid is significantly low. Power Grids should be furnished with large rechargeable batteries for energy supply through the nighttime, requiring additional recharging of circuits. This is typically very luxurious and raises the cost of production meaningfully. Therefore, the direction and position of the wind-powered sensor nodes, such as the location of the solar cell, affect the energy consumption and causes various energy harvesting rates in the Power Grid. Also, severe weather disorders like storms may damage the solar panels and turbine blades. But, the advanced control, communication, and sensing functionalities may help the addition of such recurrent power resources in the Power Grid.

1.2.4 Need for Smart Grid modernization

A smart grid is "an electric grid that comprises a range of functional and energy measures with smart appliances, smart meters, renewable energy resources." So, the smart grid is an interconnection of things that produce, transmit and dispense energy. Implementation of smart Grid leads to many advantages. It is profitable and hence saves money, and also helps in preserving energy. The Smart Grid eradicates the necessity of thermal power plants and enables greater usage of renewable energy sources, such as wind, solar, and hydro.

Furthermore, the smart grid will produce new and satisfying jobs in green industries and power plants. So, potential engineers must be familiar with the role of this technology for the future development of smart grids. This technology demands organized effort and teamwork among a remarkable discipline, with communication control architectures for generation, distribution, transmission, markets, procedures and supply management (Wang, Xu, & Khanna, 2011).

Effective employment of smart grid framework increases digital information employment and augments security, productivity, and reliability of the power grid. Smart Grid optimizes the operations with complete security. It leads to the incorporation and development of demand response, energy-efficiency resources and demand-side resources. It creates an opportunity to integrate smart appliances and end-user devices. It provides appropriate information and control decisions to consumers. Fig. 1.1 shows the advantages of the Smart Grid framework is depicted in Fig. 1.9.

Advantages of Smart Grid technology

1. **Reliable power source**

   The smart grid uses several technologies that augment the fault detection and heals the communication network without the involvement of

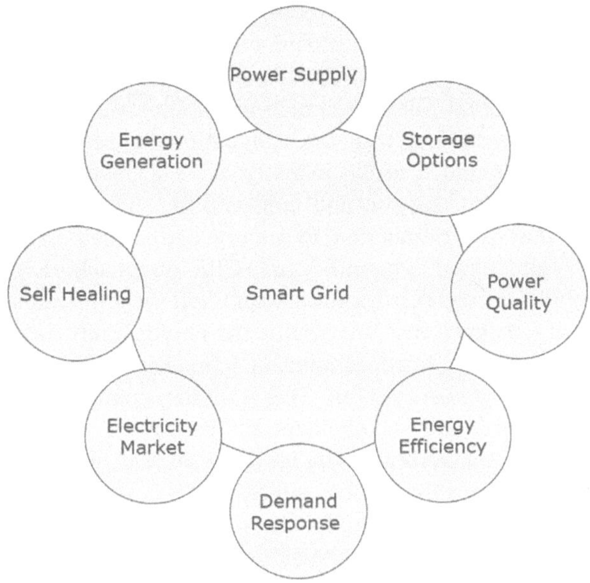

FIGURE 1.9 Benefits of Smart Grid framework.

technicians so improving consistent and reliable electricity supply and reduced susceptibility to usual disasters (Hossain, Han, & Poor, 2012).

2. **Flexibility in network topology**

   An improved structure of distribution and transmission may hold bidirectional energy flows, permitting for several options of energy production through various sources like photovoltaic panels on the roofs of buildings. The usage of fuel cells, wind turbines, charging of the batteries in electric vehicles, and pumped hydroelectric power where traditional grids support the one-way stream of electric power.

3. **Maintainable infrastructure**

   Because of the improved agility of the smart grid, it allows communication from different renewable energy resources like solar wind power and solar energy. The present network structure is not designed to support many disseminated feed-in points. Smart grid technology is an essential condition for huge amounts of renewable energy because it withstands many power fluctuations that will arise due to atmospheric conditions.

4. **Market demand**

   The smart grid permits orderly communication among consumers and suppliers to be more sophisticated and flexible in the operational policies (Gao, Xiao, Liu, Liang, & Chen, 2012). The critical loads should pay more prices, and users can become more calculated when using electric power. Domestic utilization such as refrigerators, air- conditioners, heat pumps and heat banks will hunt for minimizing energy cost.

The SG was developed as a technology to face these tasks. Smart Grids empower the 4th phase of the manufacturing and industrial revolution, termed Smart Grid Industry 4.0 (SGI 4.0). Communications and Information Technologies play a vital role in cumulating stability, reliability, and competence in SGI 4.0, compared with the conventional grid. The revolution of industry 4.0 is not bounded to the industry but also involves the whole life cycle of the generation, supply, and consumption. The computerization of the life cycle, that is, "production to supply and supply to user," can be attained by applying cloud computing, IoT, CPS, and machine learning.

The self-optimization, self-cognition, and self-customization of the SGI 4.0 enable diverged functional systems of the energy grid towards functioning autonomously, with a lesser amount of human involvement for first-rate and quality electric power. So, in SGI 4.0, the communication systems, employees, customers, and stakeholders are connected, communicated and work closely. This collaboration will be done through the autonomous and continuous exchange of information, which builds a flexible energy generation framework that can significantly improve power efficacy and keenness in power industries. This causes a change in the traditional electricity generation techniques and guides for further improvements in future terms of unpredictable external events and internal complexity. But, the autonomous multifunctional equipment in a high-scale smart grid originates a large quantity of unclear data. The traditional information exchange approaches are not suitable for handling this huge data, which intensely increases the system's complexity and leads to uncertainties because of specific events. Instead, the communication prerequisites in the SGI 4.0 are increased concerning data rates, latency, bandwidth, scalability, efficiency, flexibility and reliability.

1.2.4.1 Smart Grid architecture

Designing a strong communications network structure is very significant for SGI 4.0 to achieve efficient and reliable operation (Lamb & Agarwal, 2020). There are several SG architectural models, and they employ a multilayered structure with three main layers:

1. Power systems layer: This layer aims to deliver and generate electrical power to the consumers, like a conventional power grid.
2. Communications layer: This layer provides interconnectedness among fundamental system components utilizing information from end-user interfaces and sensors to communicate them to data centers and vice-versa.
3. Applications layer: Information is processed to deliver control and monitoring messages, along with generating the information in this layer for applications like automatic meter reading, demand management, and detecting misuse or fraud. These three layers are demonstrated in Fig. 1.2.

Smart Grids depend on the usage of many networks that differ in location and size, like a Home Area Network (HAN) that communicates with devices and smart appliances to an in-house smart meter. They have a smaller range as well as reliably communicate with lower data rates. Therefore, they reduce implementation charges and power consumption. Building Area Networks (BANs) are like a HAN but encompass bigger constructions and consist of many small networks. An Industry Area Network (IAN) is like a BAN that is extra complex and concerned for industries. Neighborhood area Network (NAN) is accountable for communication connections among WANs to HANs, BANs, and IANs and aggregates data from many smart meters (Kuzlu, Pipattanasomporn, & Rahman, 2014). NANs require higher data due to the larger scope. NANs use a Wide Area Network (WAN) for sending the electricity statements to the control center. Wide Areas Networks need high data rates and lengthy distance coverage. Here optical fiber networks are generally employed as a communications medium is depicted in Fig. 1.10.

1.2.4.2 Smart Grid components

Smart Grids are composite structures that contain several components and ICTs. The main technologies that establish a component in a Smart Grid system are Cyber-Physical Systems, Internet of Things, Data Transmission Technologies (Zigbee, LPWAN, Cellular networks), Cloud Computing, Network management technologies like SDN. Smart grid systems are much sophisticated because they syndicate many connected components. These components and the emergent nature of smart grids have led to numerous

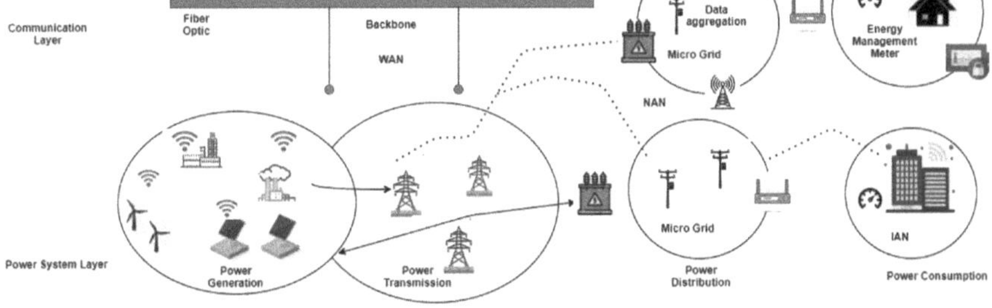

FIGURE 1.10 Smart Grid architecture.

challenges which should be reported to guarantee a scalable, robust, and reliable grid.

1.2.5 Energy sustainability in Smart Grids

1.2.5.1 Energy tolerance

Smart Grids augment the capacity of recurrent renewable generation in power systems. Smart Grids boast energy-savings utilizing:

1. Improved effectiveness of energy generation resources by enhancing scheduling technologies that have been previously widely used in renewable energy generation like solar, wind, and distributed generation.
2. New answers to generate scheduling problems in power systems with a huge grid-connected wind energy capacity.
3. Techniques to design an extreme power point chasing manager for photovoltaic systems for scheduling of Proportional-Integral-Derivative type controller using an adaptive fuzzy gain in scheduling.
4. The loss of power lines by improving reactive power compensation by developing high-energy and long-distance transmission systems.
5. Improved utilization efficiency of electrical equipment using demand response protocol which makes it possible for incentive mechanism to change ordinary electric utility.
6. The Smart Grid augments the power-consuming efficiency for the users. It provides different forms of essential data for electric energy users, like historical electricity consumption, carbon dioxide emission from the consumption of electric power, immediate demand, humidity, environment temperature.

1.2.6 Communication technologies and challenges in the implementation of Smart Grids

In Smart Grids, various Quality of Service requirements is achieved through ICTs. Allowing Smart Grid applications results in growing information quantity units from Megabyte to Gigabyte and Terabyte to Petabyte. Here we discuss major technologies that establish the composition of the SG system.

1.2.6.1 Key components of Smart Grids in the context of industry 4.0

Smart Grids are multifaceted systems that consist of various components.

1. **Internet of things**

   A communication network that interconnects device objects to the Internet through a protocol for management, communication, continuous monitoring, identification and information tracking among many smart objects and devices is said to be an Internet of things (IoT). IoT is considered a revolutionary technology that enhances conventional communication systems' functionality and helps them discover final solutions to major applications in various areas like manufacturing, logistics, healthcare, military retail, etc. In SGs, the technologies like IoT have a crucial role in enabling the data transfer between the components of SG effectively. The major application of IoT systems is Power reduction in a Smart Grid. In Smart Grids, to attain effective connectivity between applications and users, some major components are essential: smart energy meters, sensors, network interfaces, substation feeders, grid monitoring controls, smart inverters for solar energy, etc. These elements together operate to hand over information in real-time precisely.

2. **Cyber-physical systems**

   A Cyber-Physical System (CPS) integrates physical elements and cyber actively. A perfect CPS contains networking tools, computational systems, and physical devices like actuators and sensors. The technical problems of SG are monitored and controlled by controllers utilizing communicating with these sensors through a communications channel. There is a possibility to record the complete Information and position of the smart Grid, like its working status. The sensors direct complete pertinent information to these controllers for further future actions. The CPS technologies permit the SGs to achieve real-time measurement and analysis to progress in safety, cost reduction, power consumption and decision-making capability.

3. **Cloud Computing**

   Cloud Computing offers applications and services. With data processing and storage capacity over the Internet, Cloud Computing collects resources to reduce the necessity of devoted CPS and allows for advanced levels of computerization and automation. It offers three main benefits: (1) Cost reduction of the infrastructure. (2) Outsourcing and maintenance to lower both cost and risk. (3) Ease of Implementation and Scalability.

1.2.6.2 Communication technologies in SGI 4.0

Information communication and transmission are vital parts of the SG operation; therefore, it employs various communication technologies that are not offered for SG, but they are programmed for many ICT applications. The grid should receive and send data from smart meters and transmit them to remote distances for processing (Sendin et al., 2015). The SG must possess appropriate communication with traditional power grids utilizing integration technologies. The most current communication technologies frequently employed in the smart grid are described further.

Data transmission technologies such as LPWAN, Fiber, and Zigbee are major wireless communication technologies that are typically desired due to their respective ease of implementation; they are also susceptible to intrusion and need a continuous battery replacement or power charging. Cellular networks are used for mobile devices. The 3G and 4G technologies are broadly employed in cellular communication, but 5G, the fifth-generation cellular technology, is further favorable for future SG applications. It was invented to bring higher data rates along with the least message latency delay. 5G is an evolving technology with high potential for upcoming advancements in the coming years, where it is anticipated to develop additional adaptability for SG metering applications.

Zigbee is a wireless technology usually used in **wireless sensor networks**, which has been employed in smart meter systems. Zigbee products employ a relay system for transmitting communication messages from one another; it cares about many network topologies. It is also a low-cost, low-energy solution, but it is difficult to show lower bit rates, which is satisfactory. At the same time, smart meters communicate with **HANs**, which are low-range applications. **Low Power Wide Area Network (LPWAN)** protocols can communicate and effectively send information at nominal data rates. For lengthy distances, the LPWAN protocols **Sigfox and** LoRaWAN are used.

LoRaWAN is an extremely scalable WAN protocol that permits a 2-way communication among smart products, and its range is 2–5 km in urban zones and up to 15 km in suburban parts, with a data rate of 0.3–50 kbps. This protocol was exactly designed to let several applications and IoT devices communicate over wide distances.

1.2.6.2.1 Integration technology

There are diverged protocols designed for conventional energy systems that offer malleability to employ SGs let them integrate with present power grids, like a legacy communication protocol, IEC 61850 that supports integration, and DNP3(Distributed Network Protocol).

1.2.6.2.2 Network management technologies

The smart grid prerequisites of reliability and high speeds have created the need for protocol management technologies to manage configuration such as persisting effectiveness in the grid. A new networking model for centralized control of the network is through an SDN controller device, which permits the local networking appliances to finalize routing and forwarding choices earlier in a flexible way, depriving them of manual configuration. Engaging SDNs in SGs makes them further and more effective in data management and provides the ability to respond to disasters or attacks, making it further proficient in preserving its essential functionality. The other two qualities which improve grid management are computerization and virtualization. Computerization is an automation model in which functionality is deployed on a software level other than hardware, making it flexible, reconfigurable for new additions and interactions with various technologies. Virtualization increases the ease of creating virtual network components (resources, hardware) to eradicate the requirement for altering the hardware components where both concepts are achieved through cloud services.

1.2.6.3 SGI 4.0 challenges

Smart Grids experience more challenges that should be considered to guarantee a reliable and strong grid.

1.2.6.3.1 Communication system challenges

Researchers found that communication system challenges have three main issues, data transmission rates, interference, and the need for common standards.

1. Interference is reasoned by home objects' signals colliding with emissions from the grid or smart meters. Interference can be addressed using the channel switching technique and interference detection.
2. Grid standards are essential to offer a structure for every Grid component to work together efficiently. There are many efforts by various organizations, like American National Standards Institute and IEEE, to develop such standards.
3. Information about transmission rates may be accomplished by selecting the right protocol for all applications.

1.2.6.3.2 Big data challenges

Big-Data is a large volume of data delivered by contemporary processing power and information systems needed to store and analyze the data. It introduces a huge number of challenges that ought to be considered. There are some of the challenges which are faced by smart grid systems regarding Big Data.

1. **Real-time applications:** Smart Grids must process the information and execute amendments in a near real-time operation. It is suggested that some of the methodologies advanced for IoT applications also be practically employed to the SG. As an example, algorithms for prediction on house appliances are used for identifying the levels of data from a smart meter at random times in a day and helps in allocating processing sources simultaneously.
2. **Heterogeneous Data:** The Smart Grid obtains data from diverged sources in different formats. Examples include monitoring data, usage data, error messages, capacity levels, metadata, authentication messages, etc. This information is generated from divergent resources such as sensors, meters, stations, actuators, historical data, mobile applications, smart

home devices, etc. This phenomenon is known as heterogeneous data, which means the grid has to hold unstructured, semi-structured, structured information simultaneously. Numerous techniques are available to alleviate the problem, like the development of consistent software solutions which integrate various types of information from different devices.

3. **Data compression and visualization**: The Grid collects data and needs storage so that it also offers valuable analytical information; for that, it requires extra processing. Visualization techniques can be used to present the data understandably, that is, bar graphs and pie charts. The correct visualization technique and presentation of the knowledge correctly is a hard task that requires to be dealt with care. Both visualization and compression are required for more investigation and development of inventing consistent methods.

1.2.6.3.3 Challenges in cloud computing

Cloud computing technology presents a resourceful solution for processing smart grid information, yet few reliability matters exist. Safety apprehensions are owing to the nonexistence of reliable strategies regarding data location that may compromise security. The next issue is the lack of proper planning for recovery from disaster, which compromises availability. Other concerns contain international laws concerning error management, data stored in various countries, and data rates. Other issues are relating to Cost, QoS (quality of service) and compatibility. Naturally, these challenges should be dealt with by carefully selecting cloud providers to confirm that international laws are held, and the service quality is preserved.

1.2.6.3.4 Security challenges

As SG Systems are cyber-physical, they become susceptible to cyber threats that may disturb physical devices, causing security attacks very dangerous and extremely impactful. There are many types of attacks that exploit devices (sensors, smart meters, network devices etc.), and they target system building layers (application layer, network layer, communication layer etc.), or they disrupt security and privacy objectives (confidentiality, availability, integrity, etc.). Few attacks that researchers recognize are Denial of service (DoS) attacks utilizing Jamming or flooding at any level of the grid.

1.2.6.3.5 Main attacks on the grid

1. **DoS attacks:** DoS attack denotes a huge number of outbreaks which makes a system to develop incapable of providing services offered by it. This is threatening to the grid owing to its real-time operations. DoS attacks remain difficult to notice since they seem to be masked as genuine traffic. Also, such attacks may be derived from different bases, so service providers can't resist these attacks. To lessen the influence of these

attacks, the growth of strong cyber systems capable of surviving many persistent attacks is significant. An alternative technique that may be employed to decrease Distributed DoS attacks is I.P. fast hopping, which hides the I.P. address of a server or a service to make it difficult to be targeted by opponents.

2. **Data injection:** Injection attacks arise after a malevolent entity changes codes and table entries in the database to reveal valuable information. Smart Grids are susceptible to such an attack due to different devices linked to them, offering access points for enemies. Furthermore, these attacks may have a huge variety of consequences on energy systems, from troublemaking the real-time nature of the grid for making fraud by operating smart meter power pricing and readings. Data injection assaults are difficult to notice, and it is healthier to avoid them by employing smart security systems with committed validation to lessen the likelihood of these outbreaks.
3. **Privacy and Security:** Smart Grid systems gather a huge amount of consumer data, like payment data, location, usage of power and also preferences. This information will be maliciously used to track the users. A user's power usage trends can be predicted, such as when he leaves the house, therefore allowing robbers to target his house. So, protecting users' information and privacy are more critical. Several methods may be employed to reserve security and privacy in SGs, like masking sources, aggregating and encoding data through different methods that separate users from valuable data.
4. **Insider attacks:** These attacks committed by enemies with authenticated admission to the system are very hazardous because conventional security and privacy measures like passwords, firewalls may not be able to prevent them from triggering harm. Few hiring exercises such as background checks may help in reducing these insider threats. Many technological resolutions are known as anomaly detection systems that can find any erratic and abnormal behavior through authenticated access control through software solutions and gateway devices.

1.3 Smart Grids and Cyber security in Industry 4.0

The new advances in innovation, the expanded dependence on electrical energy and the development of the fourth modern upset (Industry 4.0) were all elements in the expanded requirement for keen, proficient and solid energy frameworks. This presented the idea of the SG. A Smart Grid is a possible swap for more established force matrices, equipped for adjusting and disseminating energy dependent on request. Smart Grid frameworks are intricate. They join different segments and have high prerequisites for constant, dependable activity. This chapter endeavors to outline Smart Grid frameworks by illustrating Smart Grid engineering and other components. It too presents correspondence advancements, mix and organization the executive's apparatuses associated with Smart Grid frameworks.

In the past many years, power grids were improved by staying up with mechanical, communal headways and radically increasing the electrical supply interest. Insights (Lu, De, & Song, 2010) showed that from 1950 to 2008, energy creation and utilization in the U.S. increment roughly two and multiple times, individually. Specifically, people in general/business administrations, industry, and neighborhoods are the most requesting U.S. power regions in 2008. The one significant test is to proficiently deal with an assortment of energy assets, including conventional petroleum products obtained (coal, petrol, gas) environmentally friendly power assets (e.g., solar-based and hydro-based) (Huang, Crow, Heydt, & Zheng, 2011). Then, the National Institute of Standards and Technology (NIST) carried out public endeavors to foster the cutting edge smart grids framework, ordinarily alluded to as the Smart Grid (Office of the National Coordinator for Smart Grid Interoperability, 2010). Contrasted and inheritance electrical frameworks, the Smart Grid is imagined coordinating fast, entirely. Two-way correspondence innovations (Aggarwal, Kunta, & Verma, 2010; Choi, Lee, Lee, & Ahn, 2008; Gungor & Lambert, 2006; Mohagheghi, Stoupis, & Wang, 2009; Sui, Wang, Lu, & Lee, 2009; Zhou, Guo, & Qin, 2010) into a massive number of grids types of gear to build up a dynamic and intelligent foundation with new energy the executives' capacities, for example, progressed metering foundation (AMI) (LeMay, Nelli, Gross, & Gunter, 2008) and request reaction (Metke & Ekl, 2010).

Nonetheless, a hefty dependence on data unavoidably gives up the Smart Grid to privacy vulnerabilities related to interchanges and systems administration systems. Smart Grid indeed builds the danger of giving and taking solid and secure grids system activity, which, regardless, is a definitive goal of the Smart Grid. For instance, it has been shown (Guidelines for smart grid cybersecurity, 2010) that possible organization interruption by foes may prompt an assortment of severe outcomes in the Smart Grid, from client data spillage to a course of disappointments, like colossal power outage and destruction of infrastructures. The Smart Grid is a redesign that will add numerous new functionalities to meet clients' unique force prerequisites—refreshing a framework as perplexing as the electrical force network can bring new security vulnerabilities into the framework. This record presents the business related to Smart Grid digital protection. The work surveyed is isolated into five classes that make up various parts of the Smart Grid: Process Control System (PCS) Security, Smart Meter Security, Power System State Estimation Security, Smart Grid Communication Protocol Security, and Smart Grid Simulation for Security Analysis. The Smart Grid is a hugely complex system. It is anything but a great deal of digital protection configuration work.

The Smart Grid is considered as a progressive and developmental system of existing grid matrices. The main thing is with the joining of existing grid processing and correspondence innovations, the Smart Grid is required to enormously upgrade the effectiveness and unwavering quality of future grid systems with environmentally friendly power assets, just as circulated knowledge and request reaction. Alongside the subtle highlights of the Smart Grid, digital protection arises to be an essential issue since many electronic gadgets are associated through correspondence networks all through important grid offices, which promptly affects the unwavering quality of an extensive system. Smart Grid has to consider the digital protection issues for the Smart Grid is primary importance. Explicitly we center around exploring and talking about security prerequisites, network issues, assault countermeasures, secure correspondence conventions and models in the Smart Grid. Plan to give a profound comprehension of security vulnerabilities and arrangements in the Smart Grid and shed light on future assessment headings for the importance of Cyber security in Smart.

1.3.1 Smart Grids and security issues: a comprehensive overview and its classifications

Having distinguished different dangers and difficulties by the framework, the risks should be obvious. An incorporating approach towards the functional security arrangements should be laid and followed—consequently, the need to study the wellsprings of dangers. The dangers can be seen according to numerous viewpoints, as introduced in Fig. 1.3. The order concentrated in this work depends on the specialized and non-specialized view. The recognized dangers are gathered dependent on different distinguished sources, as portrayed in Fig. 1.4. These classifications can also be modified as the figure has not provided a particular order. A model opened for suitable change and renamed into conceivable, more modest units for very much organized preventive and remedial measures—Technical and Non-technical sources of smart grids threats.

Such a lot of exploration endeavors are being set up for improved security and versatility of the Smart Grid by scientists worldwide utilizing different strategies to check the difficulties presented in various structures. Strangely, past works as they were looking to resolve an issue or give an expansive depiction with no revealed accentuation to wellsprings of dangers. Some related chips away at Smart Grid security pointed toward proffering answers for some recognized threats are discussed below figure Fig. 1.11.

1.3.2 Data privacy and security challenges in Smart Grids

Security is an essential component of any innovation identified with delicate resources, and new mechanical progressions power associations to confront

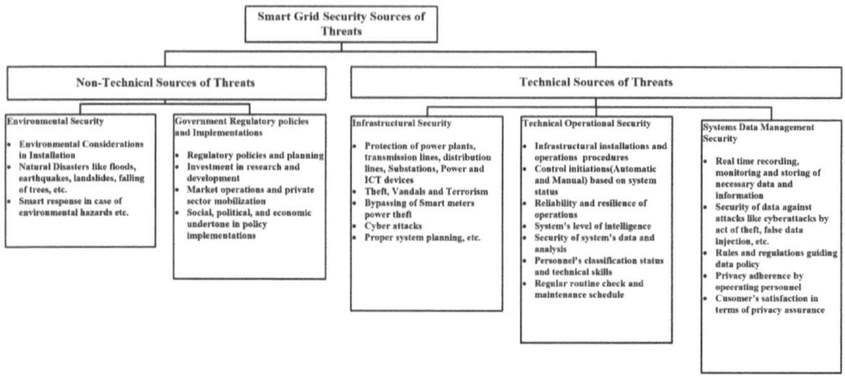

FIGURE 1.11 Taxonomy of security in Smart Grid.

new security concerns and dangers. When gadgets like keen meters are made safer, their business esteem increments since associations can react powerfully and with certainty to new market openings or evolving conditions. This considers further developed proficiency, diminished expenses, and further developed progression of administrations. Robust security for the grid can have an enormously positive effect, as an absence of safety straightforwardly undermines the wellbeing of general society and those who are influenced by framework disappointments. For instance, there is a worldwide exertion to change the electric framework to a computerized foundation. A crucial move like this is viewed as essential to stay aware of energy and utility functional requests while keeping up with or working on the primary concern. A computerized foundation can be more adaptable and dynamic; as it may, such an organization may uncover the essential foundation to extra security concerns, such as unapproved access and digital assaults.

Security can't be characterized. The fundamental meaning of security would be "the option to be left alone." Security ought not to be mistaken for privacy. Privacy of data is data that can be gotten to exclusively by a few who has the authority. Concerning the smart grid network, protection implies thinking about clients' rights, qualities, and interests (like their own data, digital marks or signatures, and so forth). The information utilized by the smart grid could be used to abuse people (Bekara, 2014). A privacy impact assessment (PIA) is being used for deciding the protection, privacy, and security dangers that emerge because of the assortment, use, and exposure of individual data.

1.3.3 Built-in security architecture and IoT conceptual model for Smart Grids

Internet of things makes the Internet the next step of development. IoT made life simpler, mechanized, and convenient by getting the entire world into the hand through calculation and correspondence capacities. With the smart grid's progression and its parts, innovation was expected to associate these segments, dependable and more competent effectively. IoT has a guarantee to fulfill this load of qualities bringing the smart grid into the new period. Be that as it may, some genuine security concerns arose with this innovation, which incorporates pantomime, information altering, exaggerating, approval, protection issue, and digital assault (Patel, 2016). Scientists are doing concentrate on managing these issues. IoT-based savvy networks should have administrations like confirmation, secrecy, client protection and information honesty to avoid any security risk (Uribe-Pérez et al., 2016). Availability that IoT gives to the client improves their experience and proficiency. It is anything but adaptable and simple cooperation with the framework to decrease cost by diagnostics and neighborhood-wide meter perusing ability (Gao et al., 2012). To put it plainly, it makes a smart grid smarter.

1.4 Conclusions and future work

Smart Grids are superior to conventional inheritance power matrices as far as capability and efficiency, as the Smart Grid is harmless to the ecosystem. It utilizes a ton of inexhaustible wellsprings of energy and principal. It is safer than the customary force matrix. Besides, the examination recommended privacy advantages and vulnerabilities against the Smart Grid. The benefits of utilizing a Smart Grid in the general viewpoint are anything but a more extensive scope of safety with different strategies and methods to conquer a portion of the digital assault issues. While directing the examination, the various articles have recommended the security advantages and vulnerabilities of Smart Grids. Pretty much every exploration article proposed that compromising soft spot for Smart Grids would be the Denial-of-Service assault. Since Smart Grids are the development of the organization and assaulting the organization would disable the Smart Grid. The Smart Grid will ensure the availability of numerous layers of safety. An ideal answer for the security angles would utilize the Virtual Private Network. And a lot of development is necessary for future exploration for various angles in various spaces of the smart grid framework. Smart Grids remember anticipation, power stream enhancement, correspondence, miniature network coordination, request and energy the executive's framework, conformance of norms for interoperability, adaptability, sensitive components, information encryption and above all robotization of age, transmission and circulation.

Also, closing this exploration, mindfulness identified with digital assault in Smart Grids is significant. The client ought to know about the dangers identified with the Smart Grid and moderate them by doing different danger appraisals and contextual analyses to secure the Smart Grid against various sorts of digital assault. Moreover, the examination addressed privacy difficulties identified with the Smart Grid. Smart Grid challenges are those different gadgets associated with huge geological region organizations. The most important test is to get these gadgets over the more robust framework. Blockchain innovation could assist with settling security issues by giving a familiar and encoded record that is unchanging to changes made by noxious hubs or aggressors. It can likewise be used to check personalities and approve access by putting away and recording exchanges in the permanent record and make information trades between appropriated contraptions smooth and cost-productive. The PC network conventions should be changed by the current stance of correspondence, just as giving complex encryption strategies and offering security countermeasures. Hence, it will protect against developed digital assaults.

References

Aggarwal, Kunta, S., & Verma, P. K. (2010). A proposed communications infrastructure for the smart grid. In *Proceeding of innovative smart grid technologies conference Europe (ISGT).*

Alsaidan, I., Alanazi, A., Gao, W., Wu, H., & Khodaei, A. (2017). State-of-the-art in microgrid-integrated distributed energy storage sizing. *Energies*, *10*, 1421.

Amarsingh, A. A., Latchman, H. A., & Yang, D. (2014). Narrowband power line communications: Enabling the Smart Grid. *IEEE Potentials*, *33*, 16–21.

Bekara, C. (2014). Security issues and challenges for the IoT-based smart grid. *Procedia Computer Science*, *34*, 532–537.

Choi, T.-I., Lee, K. Y., Lee, D. R., & Ahn, J. K. (2008). Communication system for distribution automation using CDMA. *IEEE Transactions on Power Delivery*, *23*, 650–656.

D.U.S.C.B. Patel (2016). IoT enabled Smart Grid, IJSRD, National Conference on ICT & IoT, January 2016.

Gao, J., Xiao, Y., Liu, J., Liang, W., & Chen, C. L. P. (2012). A survey of communication/networking in SmartGrids. *FutureGeneration Computer Systems*, *28*(2), 391–404.

Gelazanskas, L., & Gamage, K. A. (2014). Demand side management in Smart Grid: A review and proposals for future direction. *Sustainable Cities and Society*, *11*, 22–30.

Guidelines for smart grid cybersecurity. (2010). vol. 2. Privacy and smart grid. http://csrc.nist.gov/publications/nistir/ir7628/nistir-7628vol2.

Gungor, V. C., & Lambert, F. C. (2006). A survey on communication networks for electric system automation. *Computer Networks*, 877–897.

Hossain, E., Han, Z., & Poor, H. V. (2012). *Smart Grid communications and networking*. Cambridge, United Kingdom: Cambridge University Press.

Huang, M., Crow, G., Heydt, J., & Zheng, S. (2011). Dale, The future renewable electric energy delivery and management (FREEDM) systems: the energy internet. *Proceedings of the IEEE*, *1*, 133–148.

Jerin, A. R. A., Prabaharan, N., Kumar, N. M., Palanisamy, K., Umashankar, S., & Siano, P. (2018). *Smart Grid and power quality issues. Hybrid-renewable energy systems in microgrids* (pp. 195–202). Amsterdam, The Netherlands: Elsevier B.V.

Kuzlu, M., Pipattanasomporn, M., & Rahman, S. (2014). Communication network requirements for major Smart Grid applications in HAN, NAN and WAN. *Computer Networks.*, *67*, 74–88.

Lamb, M., & Agarwal, R. (2020). Deploying digital assets: Natural gas utilities using Smart Grid technologies to modernize infrastructure. *National Gas & Electric.*, *13*, 5739.

LeMay, M., Nelli, R., Gross, G., & Gunter, C. A. (2008). An integrated architecture for demand response communications and control. In *Proceeding of 41st Hawaii International Conference on System Sciences (HICSS' 08).*

Lu, G., De, D., & Song, W. -Z. (2010). Smart Grid Lab: A laboratory-based smart grid testbed. In *Proceeding of IEEE conference on smart grid communications*.

Metke, A. R., & Ekl, R. L. (2010). Smart grid security technology. In *Proceeding of innovative smart grid technologies conference Europe (ISGT).*

Mohagheghi, S., Stoupis, J., & Wang, Z. (2009). Communication protocols and networks for power systems—Current status and future trends. In *Proceeding of power systems conference and exposition (PES' 09).*

Office of the National Coordinator for Smart Grid Interoperability. (2010). *NIST framework and roadmap for smart grid interoperability standards, release 1.0.* NIST Special Publication 1108, pp. 1–145.

Portelinha Júnior, F. M., Zambroni de Souza, A. C., Castilla, M., Queiroz Oliveira, D., & Ribeiro, P. F. (2017). Control strategies for improving energy efficiency and reliability in autonomous microgrids with communication constraints. *Energies*, *10*, 1443.

Sendin, A., Arzuaga, T., Urrutia, I., Berganza, I., Fernandez, A., Marron, L., ... Arzuaga, L. (2015). Adaptation of powerline communications based smart metering deployments to the requirements of smart grids. *Energies*, *8*, 13481–13507.

Sui, H., Wang, H., Lu, M.-S., & Lee, W.-J. (2009). An AMI system for the deregulated electricity markets. *IEEE Transactions on Industry Applications*, *45*(6), 2104–2108.

Uribe-Pérez, N., Hernández, L., de la Vega, D., & Angulo, I. (2016). State of the art and trends review of smart metering in electricity grids. *Applied Sciences.*, *6*, 68.

Wang, W., Xu, Y., & Khanna, M. (2011). A survey on the communication architectures in smart grid. *Computer Networks*, *55*(15), 3604–3629.

Zhou, H. J., Guo, C. X., & Qin, J. (2010). Efficient application of GPRS and CDMA networks in SCADA system. In *Proceeding of IEEE power and energy society general meeting (PES' 10).*

Chapter 2

Recent regulatory control changes related to energy storage in Asia: impact on the business cases

T. Ramathulasi[1], T. Anitha[2], P. Saraswathi[3] and A. Anupama[2]

[1]*Department of Computer Science and Engineering, Mother Theresa Institute of Engineering & Technology, Palamaner, Chittoor, India,* [2]*Department of Computer Science and Engineering, Anil Neerukonda Institute of Technology and Sciences, Visakhapatnam, India,* [3]*Department of Computer Science and Engineering, GITAM Institute of Technology, Visakhapatnam, India*

2.1 Introduction

Concerns about global climate change need to be addressed for the adoption of mechanisms needed to decarbonize energy in the energy renewable process in a solar and wind-based renewable energy (RES) source system (Narayanan, Mets, Strobbe, & Develder, 2019). For example, through Asian policies, we have set a strong challenge to obtain a share of the total energy consumed by RES by 2050 (Hübler & Löschel, 2013). Reflecting this, surveys show that solar and wind energy efficiency have gradually improved over time Similarly, we hope to get even better in the future. The main problem posed by grid operators is the network imbalance caused by the intrusion of various resources such as pure wind turbines in the power system and the intermittent change of RES. It also leads to a decrease in system efficiency and an increase in costs and reserve capacity (Brouwer, van den Broek, Seebregts, & Faaiji, 2015). To improve and ensure the power system with a large share of RES in terms of stability, it is necessary to adopt methods such as setting up and maintaining capacity markets, smart grid (Hossain, et al., 2016) programs, etc. The focus is on introducing identified energy reserves (ES) to maximize reserves, meet peak demand over time, and resolve interruptions in RES (Peker, Kocaman, & Kara, 2018). It is expected that the ES fuel system, which is three times the global capacity, will be required by 2030 to double its RES share (IRENA, 2017). To this end,

Sustainable Networks in Smart Grid. DOI: https://doi.org/10.1016/B978-0-323-85626-3.00007-7

studies on environmental performance, economics, and technology for ES systems and their compatibility with many grid applications (Ding, et al., 2019) are analyzed. BESS (Battery Energy Storage System) is one of the most popular types of systems currently available in ES. The reasons for this are a reduction in its capital expenditure in the coming years and improved battery materials, frequency regulation. In addition, grid-connected energy arbitrage (Kocer, Cengiz, Gezer, Gunes, & Cinar, 2019), peak shaving, and reserve capacity (Gailani, Al-Greer, Short, & Crosbie, 2020) have been improved in power networks. Bloomberg estimates that the BESS price will be $70/kWh by 2030, which is expected to be in line with grid applications and high investment ES market growth. Therefore we have discussed in this paper how to analyze regulatory changes in the markets with different ES-compliant work related to BESS.

Acceleration in energy reserves by renewables has been in place for over a decade. Purification and greater penetration of energy are part of the transformational process of global renewable energy as key components of the missing puzzle. In this context, new projects are sprouting for energy reserves in the markets, while developing countries around the world are trying to improve reserves in industry. The main reason for this is the significant drop in energy storage costs in countries around the world. In fact, (Eller & Gauntlett, 2017) estimated that we will be see an increase in energy reserves from 2 to 80 GW capacity from developing countries around the world.

The Paris Agreement represents an international movement to achieve clean energy by gaining parallel movements on climate change. From November 4, 2016, the terms of this agreement are aimed at achieving industrially lower global warming 2° C. It is hoped that investments, including those with government assistance, will largely be derived from energy renewables and storage innovations. A better reduction in storage costs can also be obtained. In 2016 investments in grid-scale battery reserves exceeded the billion-dollar index (IEA, 2017).

2.1.1 Preliminary concepts

The services required to provide various services of BESS in the electrical network are technically available for operation. Nevertheless, BESS faces large-scale disruptions in the networks of various EU systems, including large-scale energy networks, which prevent it from controlling reserves and market barriers. Analyzing the value of BESS in energy markets with the help of a model would face several obstacles under current energy law (Hamelink & Opdenakker, 2019). Researchers have said that the regulatory framework (Poyry, 2014) recommends that businesses be forced to choose which business models are currently financially viable. Base technologies are used in energy networks that do not contribute to the reduction of capital expenditure for the growth of applications. There is a need to improve the barriers to BESS regulation in commercially different ES systems markets. The modern better business case involves exploring ways to identify the factors that are needed to regulate energy in various energy networks of ES. The key to this process is to identify the ES and bring about the necessary regulatory changes. Some of the following have been adopted by Asia's energy regulator. The law primarily defines electricity storage, eliminates the need for a double network for balanced storage charges, collects reserves on sites and supports utility levy through policies, provides limited reserves to network owners, implements a simple simplified approval plan, and stores in the utility market.

In this research, we will review the benefits of the currently amended proposals for ES regulation. It also explores the impact of regulatory changes in Asia on systems of BESS business models. Control changes are considered equally in Asia as they extend the strengths (S), weaknesses (W), opportunities (O), and threats (T) of the regulatory framework proposed in the future using a "SWOT" analysis. External and internal factors are described and examined in the proposal for regulatory changes. Since this qualitative data is extracted from government and magazine reports, articles, energy regulators, this important analysis can be useful for understanding the cases surrounding the regulatory framework of ES with a clear understanding for future business. It should be noted that these current proposed regulatory changes will need energy regulators to neutralize the market business cases and impacts of BESS. Therefore all systems in the ES, including the BESS, show an equally improved increase in their use of power networks.

2.1.2 The role of energy reserves in the future

According to the report of energy storage regulations in the traditional grid system, the storage implementation rate is a major challenge. We see the relationship between decentralized power plant design and supply system as a hybrid country. Fuel storage is evident in the increasing power demand in the fuel mixture and, thus, increasing renewables and living utilities must be alive for the coming integration. The reason for this is to increase the need for a grid in the daily cyclical-demand trending solution process.

2.2 Theoretical background

2.2.1 History

Since 1970 it has been possible to develop batteries in a rechargeable manner. Beta alumina with high ionic conductivity was discovered by scientists for the Ford Motor Company (Yao & Kummer, 1967) in 1967. The solid-state electrochemistry revolution (Whittingham & Huggins, 1972) led to this discovery. This has resulted in the reversible storage of hydrogen-type Lani5, TS2, lithium, and the use of solid compounds. The reuse of NiMH and Li-ion batteries, respectively, was based on this. Forty years of successful commercialization resulted in the production of a sodium-sulfur battery. In addition, the Nihon Gaishi Kaisha battery was introduced to the Japanese market in Europe as a result of extensive research. In this battery, the reactions of sulfur and sodium are in a liquid state, and the working environment is greater than 300 C. These can also be separated in the form of a lattice or by ceramic tubing. Such batteries are used not only to power up to 200 MW but also under the typical constant loads of leveling applications. Na/S battery development in South Africa has been done with zebra cell variation, similar to the one discussed in Long Island. For example, a mixture of nickel and sodium chloride is contained by the cathode in this cell. It is present when charging nickel and sodium chloride are deposited at the cathode. Its cell advantage lies in the fact that sodium is not kept in the metal maintenance manufacturing process. It is currently being actively marketed by General Electric (GE) in Schneider, New York, for sustainable applications.

2.2.2 Evaluation

Extensive research began in the early 1970s on lithium batteries that were reusable and operated at room temperature. Energy reserves are optimized through intercalation reactions. In chemistry, intercalation is the transformation of an atom or ion into a crystalline lattice without significant changes in that lattice or a small contraction or expansion. Stanford University (Stanford, CA) was particularly interested in research in the 1970s on the modification of the physical properties of materials with superconductivity and the incorporation of atoms or ions into host structures in their structure (Gamble, et al., 1971) Obtaining the desired features engineered by inserting electron-accepting donors into the host lattice is challenging. Superconducting temperatures were altered by the work of chief scientists at the Bell Laboratory in Exxon, New Jersey. There, they discovered that the intercalation reaction can greatly improve energy and increase interest in energy reserves. The Bell and Stanford labs continued to compete for the next 10 years. It can be considered as the golden age of research and development related to lithium batteries. As a result, all secondary lithium batteries must be mounted between two electrodes using intercalation reactions (Whittingham, 1976) for energy storage. This basic technique has been followed for almost 40 years.

2.2.3 Energy system

Hydron energy storage is pumped by a low-cost fitted system. Water is supplied upwards so that the production capacity is high in the available lakes, and downstream flow is allowed by pump turbines when peak demand is high. Today's round trip with turbines with 70% (energy out / energy in) capacity. The first hydro-equipped pump was commissioned in 1909 at the Schaofusen site in Switzerland. It can generate 1 MW of power. At Blenheim-Gilboa, New York, the system is large enough to generate more than 1.1 GW of power (NYPA, 2010) The Tennessee Value Authority (TVA, 1978), which has an even larger raccoon mountain facility, can reliably generate approximately 1.6 GW of power per day, and water can fall up to 990 feet due to the upper reservoir for turbine pumps. On top of the mountain, a 528-acre reservoir and dam with a 28-hour box time to fill is a large rock-filled TVA structure with a height of 8500 and 230 ft, respectively. Bath County, VA has the largest pump in the world with a capacity of 3 GW (Dominion, 1985) hydro facility.

2.2.4 Characteristics

With a capacity of 25 GW, there are currently about 250 locations in the United States that have energy reserves. This is only 2.5% share of the total capacity. Many of the limited areas remain where the water pumping system was installed. In addition, no more dams are being built in the United States. However, the system must be able to be buried invisibly to open up opportunities. Found in the Snowdonia Mountains of Wales near Dinorwig (Dinorwig, 1984), this system generates 288 MW of electricity. Compressed air mixed with combustible natural gas or alone is used for pumping water. The latter type served as an example of a system based in Huntorf, Germany, which has been highly successful for over 20 years. It is now being used by wind power for energy storage (Crotogino, Mohmeyer, & Scharf, 2001) Another system at McIntosh in Alabama was used by the commercial sector as an alternative to a similar system, and off-peak power was used by air compression. One-third of all-natural gas is used conventionally (PowerSoutH, 1991). The "Clean Compressed Air Energy Storage System" (CAES) has been popular in Iowa for some time. The use of off-peak power and wind does not indicate that the planned structure will be built. Restricted technology has been redone here based on the availability of caves. Therefore solutions to the supply of stocks by distribution must be sought in another way. Energy reserves are carried by potential compressed air systems and pump-equipped hydro. Flywheels, by contrast, store energy by kinetic energy. They can rotate at better speeds and store more energy. In addition, it has developed an application for smoothing the frequency of flywheels without the need for long-term storage. Carbon alloys and mixed alloy flywheels are used to achieve better, higher yields. An example of such a system is the 20-megawatt and 5-megawatt generation and storage capacity operations launched in 2011 in Stephentown, New York. It, therefore, delivers approximately 200 flywheels, up to a maximum output of maximum output. Developer Beacon's failed attempt did not succeed as Power had hoped. However, the buyer will report that the system has a costly facility. He, therefore, wants the continuation of this facility. However, it is unclear whether safety issues and large systems made possible by the flywheels of strong bunkers that are in demand will be largely feasible.

2.2.5 Existing mechanism

Currently, the economically profitable load leveling method is economically advantageous in some special cases by using batteries. The Na/S battery system in Long Island, NY was used as a 1-megawatt drive to provide power to bus compressors powered by natural gas at maximum intervals. Binghamton, NY has an 8-megawatt power supply with a Li-ion battery, where ~$23 million will be spent on it. This is the most expensive location for large installations. Exploration has also begun on Long Island for a 400-megawatt facility, where battery contact with capacitors begins. There are many other possibilities, such as fuel cells. These are not commonly used for reuse so they are described as the best. Because of this, we do not use this energy for repeated storage. However, many of the properties of fuel cells are similarly to bimetal-air batteries. These are storage systems that show some promise in the future, where their characteristics and history are considered by batteries and capacitors, respectively. Moreover, they are part of electrochemistry. Batteries and capacitors are considered energy storage devices. This is clearly shown in Fig. 2.1. Storage capacities have been compared by many systems. The energy released by additional capacitors is usually low in terms of quality. The reason for this is that the strong work is distributed by the voltage discharge state. But in this condition, the battery will have a constant output voltage. Liquid fuels, such as methanol, make it possible for fuel cells to produce limited energy, with high energy storage. Similarly, energy storage is high in Li-oxygen but less energy is produced through them. In addition, it is desirable to pair the batteries with a storage medium that has optimal capacity and response time in a stable product.

2.3 Energy storage classification

2.3.1 Definition and classification

Stored in many forms, the electrical energy in ES devices is converted back into electrical energy in many forms. The storage form, size, and discharge time factors of the main types of ES systems can be divided into their applications as described in Fig. 2.2. Depending on the type of storage we want and the application we want to develop, the required

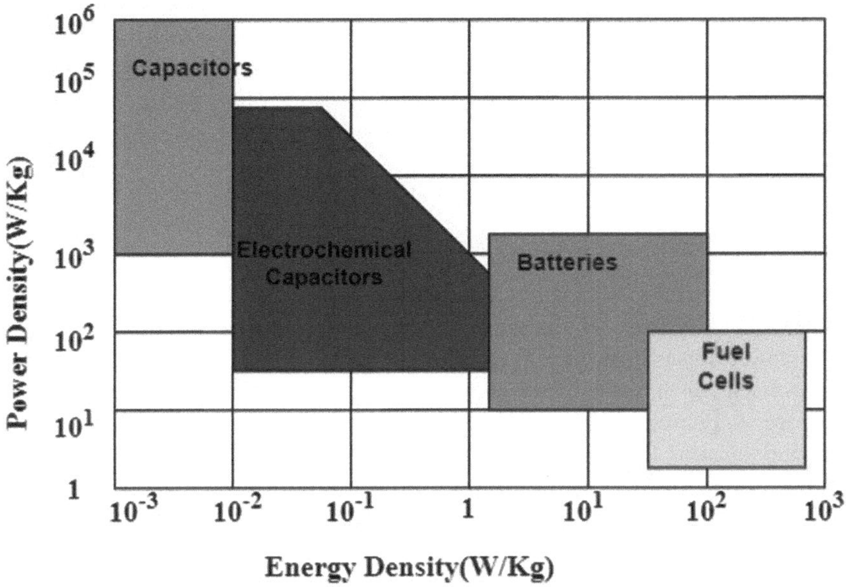

FIGURE 2.1 Power versus energy density for various energy storage systems (Kötz & Carlen, 2000).

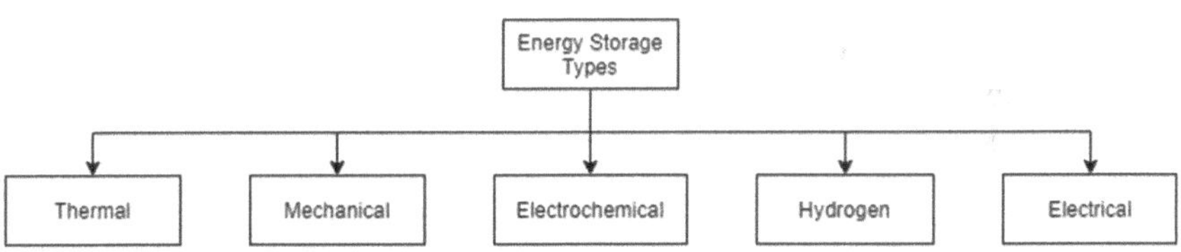

FIGURE 2.2 Classification of energy storage systems.

electrochemical, mechanical, and thermal energy, and charge time are selected accordingly. For example, heavy power management provides power to conventional production plants by Pumped Hydro Storage (PHS).

From Fig. 2.2 it can be seen that all the services in the power network can be provided with short-term BESS power distribution or transmission, supply, and bulk power storage management with the help of different types and batteries represented. In South Australia, for example, lithium-ion batteries have been used for heavy power transmission management. (Faunce, Prest, Su, Hearne, & Iacopi, 2018) BESS is used for support by network operators according to grid fidelity. They also start using services that help them calculate the value of BESS. Voltage support, barrier management, and peak shaving, which are BESS technologies in regeneration, are successfully piloted by various distribution network operators (DNOs) to develop applications for various grid applications. Among the ES power networks on the market, one should try to increase their share by considering their different applications and capabilities. ES devices can convert electrical energy into several forms that can be stored and converted back to electrical energy again.

2.3.2 Regularity changes in energy storage

The competitiveness of ES technologies is influenced by policies that promote RES expansion. In Germany, for example, wind farm operators are not encouraged to store the energy generated, as 95%–100% of the corresponding wind feed-in tariff (Fit) is paid to reduce that energy (Schermeyer, Vergara, & Fichtner, 2018). CFD (Contract for Difference), RO (Renewables Obligation), and FitzCal, CL (Climate Change Levy) have been introduced in the United Kingdom as examples to promote RES expansion to FCL policies. These tariffs are levied on the storage of electricity generated during import. However, it has been argued that rising costs in the form of tariffs should exclude employers and end-user ES energy reserves from increasing operating costs (OFGEM, 2017). The cost of the FiT levy and RO was found to be 80% (Bradbury, Laguna, & Papadopoulos, 2016) of the box cost when charging the corresponding supply and power battery. It has been proposed that FCL payments be exempted from licensing, with the new ES being redefined and modified with their features in the current era, with the main purpose being to export stored electricity to the grid instead of consuming it. However, the required payments are exempt from obtaining this generation license for FCLs, treatment of ES.

FiT has been described as a scheme used to pay for energy surplus payments and encourage customers to generate their electricity from RES with their businesses. RO works in conjunction with FIT to generate electricity through large renewable generating plants and purchase it from power suppliers. Both RO and FiT medals take their technical type as standard in determining eligibility criteria and tariff efficiency rate criteria. As a result, payments (Pearce & Slade, 2018) are made to the owner. Accreditation may be compromised during the installation of solar PV storage, which may include some of the technologies available at Fit / RO sites. For example, if a combination of both battery PV solar and metering arrangement is installed, the stored electricity payments can be received by the owner grid. According to the contractual capacity on the FiT/RO sites, permission must be granted for storage installation with the main purpose of producing and storing from renewable resources by the terms of the recent OFGEM (OFGEM, 2018). Arrangements need to be made to find out the difference between electricity meters and export-imported electricity in terms of conditions.

The basis of the CFD agreement is the acceptance of payment on (clean) electricity generated by the CFD facility. Clean power generation based on the CFD contract is done through the CFD policy and receives payment. The possibility of importing electricity from the grid to storage can be obtained if defined in the CFD facility and it is certainly not generated by the RES and will be exported at another time and lead to compromise. Both of the proposed options have been introduced by the UK Government to address this issue (BEIS, 2017). First, any storage device in the CFD facility must be registered in a separate metering unit to ensure storage independence of the CFD facility. First, the convenience of the storage device in the CFD facility is ensured by registering the storage independence separately in the metering unit. Subsequently, some special metering arrangements have to be made not only to record the electricity reserves generated in the same metering unit but also to prevent reserves. Levy Exemption Certificate (LEC) can be obtained in case the electricity is generated by RES.

Therefore there is a case where electricity is imported by CCL with the help of an ES generator device. In this case, the double CCL end users will get for industrial use. There is no proper resolution on this issue in the regulatory provision of Asia. Based on these proposed regulatory changes, it is expected that the storage CCL will not be paid in case of import of electricity. CCL is still making payments as a generator, but there is no clarity about this in the regulatory provision in Asia. Based on the details in this proposed provision, it was assumed that the storage CCL would not be paid when imported but still used as a payment generator.

2.3.3 Energy storage network charges

Network costs incurred for various power suppliers, consumers and generators will have to be met by balancing and paying network charges. Appropriate procedures for the payment of charging and balancing charges for ES, in this case, double network charges, are charged when charging and discharging due to the lack of a clear definition in EU law (Fulbright, 2022). At Asia, payments are made with a higher storage capacity, and the "Transmission Network Use of System (TNUoS)," "Balancing Services Use of System (BSUoS)" charges are transmitted in transmission. Charges paid for the delivery of services based on the "Common Distribution Charging Methodology (CDCM)" are paid only in the case of low storage capacity and distribution connected in the distribution network. However, the current variable distribution system charging system is in place (adbury, Hayling, Papadopoulos, & Heyward, 2015). Therefore the cost of setting up a storage connection on these lines is higher than the cost of setting up high voltage effective lines. OFGEM has proposed several policy solutions for a level playing field with the help of technologies appropriate

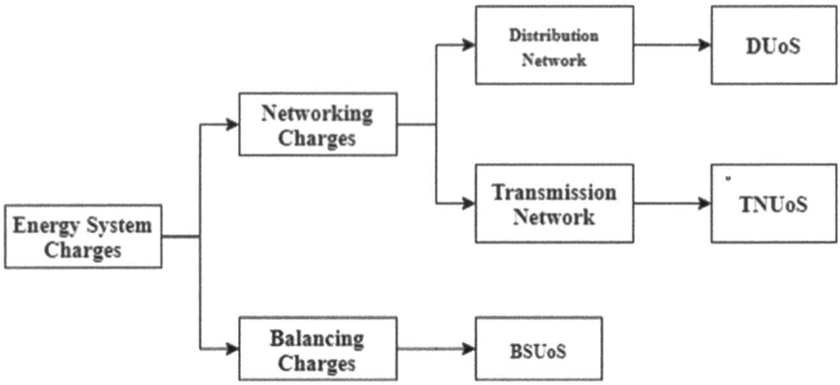

FIGURE 2.3 System charges for electrical storage system.

to this generation for these charges (Gailani, Al-Greer, Short, & Crosbie, 2020) in Fig. 2.3:

- Eliminate residual charges to meet the demand of this generation of TNUoS.
- Removal of residual charges following DUoS demand.
- Removal of residual charges by BSUoS demand.
- Newly fixed charges are implemented to cover "behind the meter" for the implementation of increased reserves.

2.4 Energy storage treatment in the market

The business approach of ES is important in the CM of the case (Khan, Verzijlbergh, Sakinci, & De Vries, 2018). In this case, the efficiency and effectiveness of the ES are called into question due to the limited energy discharge capacity required for it. Therefore ES participation in some Asian markets is limited. So-called "Unbounding" is used to illustrate the simplified and distinctive market structure used in Fig. 2.4 to streamline the competition of network operators [DNOs and transmission network operators (TNOs)] for non-network and network activities without distortion. Storage for grid reinforcement or balancing is defined as the property of this generation according to the rules subject to certain exceptions. However, there is a lack of clarity in the procedures for the proper implementation of these rules and regulations in Asian member countries amidst various exceptional circumstances. In the United Kingdom, network operators licensed under Generation licenses are exempt from storage costs with an ES capacity of up to 99 MW, as described in (Gailani, Al-Greer, Short, & Crosbie, 2020). Also, the "Talian System Operator (TERNA)" is licensed to own and operate some ES projects to gain regulated traffic on the transmission network (Graditi, Ippolito, Telaretti, & Zizzo, 2016).

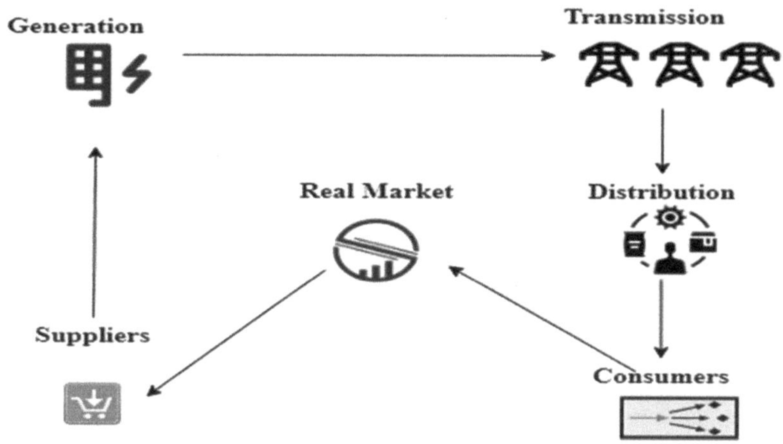

FIGURE 2.4 An example of a market structure for globalized electricity.

2.5 SWOT method analysis

Significant investment barriers and their classification have been researched as in (Castagneto Gissey, Dodds, & Radcliffe, 2018). These are mainly connected with 16 barriers and regulatory mechanisms in public attitude. Since then, regulators in Asia have dealt with various obstacles, as indicated in the previous section related to energy. However, due to the complex and ambiguous interconnection of national networks in the future, and their regulatory regulations, it is clear that there will be no better resolution in the development of ES. In an attempt to address the issue mentioned above in this paper, SWOT analysis has been used to take into account the recent regulatory regulation in the regulation of the market and thus improve the business efficiency of the future of base electricity.

SWOT analysis has been introduced as a project strategic plan and widely used tool for resolving the above criticisms and provides potential assessments as well as threats, strengths, and opportunities to traders (Huang, Fan, & Furbo, 2019). Research on the policies and technologies of national and international energy networks (Wang & Li, 2016) is of great use. SWOT analysis makes possible not only the beneficial factors caused by the proposed regulatory changes but also the harmful ones. Direct control and power regulators, which are internal factors called strengths/weaknesses in the storage business, dictate the course of business and changes. Opportunities/threats indirectly affect the ability of the warehousing business to provide regulatory changes and factors beyond the market as indirect factors. It should be noted, however, that qualitative testing, both subjective and incomplete, can be obtained by SWOT analysis. So, in this paper, they are leading the assessments in SWOT analysis with the help of recent research reports.

2.6 Results and discussion

The analysis of the general structure is described below with discussions of self-explanation and results with a better interpretation of the results (Gailani, Crosbie, Al-Greer, Short, & Dawood, 2020) in Table 2.1.

System charges by ES owners, as well as network additions, are part of the cost of exporting/importing electricity to the grid. For example, the total cost of DUoS and TNUoS is estimated to be up to 80% of the cost of transmission over the network for the ES project (Gailani, Al-Greer, Short, & Crosbie, 2020). As such, eliminating charges is an important step in the process of reducing operating costs for this ES project. The ES project has no pre-payment for the de-rating factors in this market and their participation. The full cost of the clearing was borne by the storage equipment. Section 3.4 discusses that changes in de-rating factors contributing to the declining role of BESS in this market. For example, the base for the T-4 auction deal in 2018 was 158 MW, much lower than the previous auction (Winfield, Shokrzadeh, & Jones, 2018). Fig. 2.4 shows that the number of batteries with 1- and 0.5-hour duration decreased from 91 and 40 to 2 and 17 in the auction from 2018 to 2019. In addition, the number of T-4 batteries with a life span of 1.5 hours increased from 8 to 18 from 2021 to 2022, which were also allowed to display H3 and H4 discharge batteries. The longer the storage period and the shorter the downtime, the better the results will be due to regular commitments to other services. For the battery industry, however,

TABLE 2.1 SWOT analysis of energy storage market and legislative changes in the Asia.

Aspects	Storage business case aide	Storage business case opponents
Internal	*Strengths*	*Weaknesses*
	• Getting rid of double network costs • Sites where energy storage (ES) and FCLs are colocated • Access to the grid for ES is being made easier	• ES categorization and explanation • Strengthen the requirements for unbundling. • In the capacity market, using de-rating factors • ES behind the meter will be subject to fixed tariffs
External	*Opportunities*	*Threats*
	• Promoting private sector investment in ES • Aggregation of ancillary services	• High initial investment costs • Revenue streams accessible to ES are being cannibalized

higher payouts for longer storage time were made possible by the revised regulatory instruction, thereby leading to an increase in energy and energy density Fig. 2.5.

A summary of the results obtained from the above experiment is given in Table 2.1. This has led to policy discussions on ES in Asia, implications, and the Brexit issue. We also identified the weaknesses and strengths of the ES approach as shown in Table 2.1. Some of these represented opportunity policies indirectly bring some travel issues. Reducing equipment costs in ES management, one of the main power points in Asia, has been made possible by the elimination of charges from the double network in the markets. This is done by simplifying the backup process planning permit locally and nationally for devices in the ES system. Copositioned with EO devices of RO and FiT support schemes as enabling factors. Therefore we have succeeded in identifying green electrical systems as the key enabler.

The storage definition of this generation of ES energy limits its role due to weaknesses. Income-related flows are prevented by this classification taxonomy. Network owners have been relieved from many other size barriers, such as preventing and owning ES from operating, paying FCL payments for small-scale ES, introducing fixed charges for back-to-back stocks, and identifying de-rated ES capability in CM. These barriers can erode the asset gains of the ES system. All the weaknesses mentioned here can be overcome by many possibilities. For example, to strengthen the "bundling" requirements for employers in the network, private parties can encourage investments in ES as well as better returns. It is also another possibility to allow a combination of different ES assets on a small scale to receive significant support services. Although this combined ES system is visually appealing to assets, it is important to secure the revenue streams generated by it, from the risk of cannibalism.

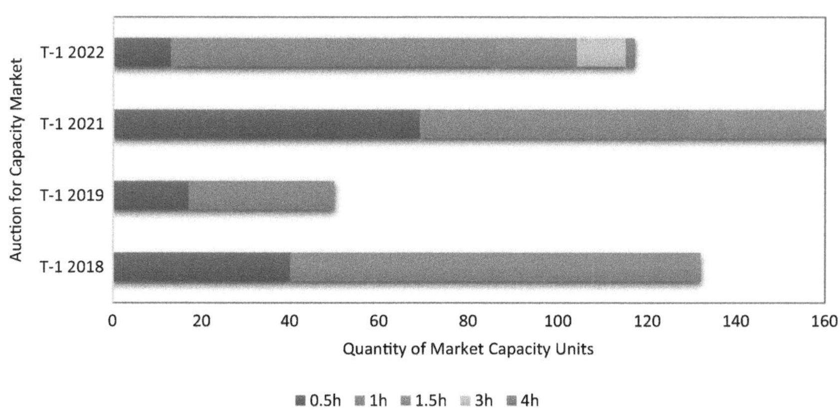

FIGURE 2.5 Capacity unit auctions in the capacity market.

Another threat is the current high capital expenditure of ES. The high capital expenditure of the current ES is another threat. The United Kingdom's exit from Asia (Brexit) could have a significant impact on SWOT factors in the ES business case study. Some studies (Lockwood, Froggatt, Wright, & Dutton, 2017) have analyzed the impact of Brexit implications on the power sector and the economic benefits lost by Internet connectors in the United Kingdom and Asia countries. Valuable services for the United Kingdom have been provided to the power network due to the operation of RESs through interconnectors. In the case of Brexit, the choice of ES and other systems was necessary to eliminate the disruption of the RES power market, which did not have an interconnector share (Mayer, Ball, Vögele, Kuckshinrichs, & Rübbelke, 2019). Therefore a recent study proves that the search for expansion of ES by Brexit in Asia should continue. Higher ES financial investments and shared interconnectors can achieve maximum power efficiency (Geske, Green, & Staffell, 2020). In the Brexit scenario, the imported batteries may not be able to meet the costly demand due to the trade arrangements in the United Kingdom and the tariffs imposed. Therefore, to satisfy the local demand and establish Giga battery factories for the grid of electric vehicles, it should be adopted in Asia. Faraday estimates that the United Kingdom will need at least eight gigafactories by 2040.

2.7 Conclusion

Nowadays, ES has been identified as better technology to prevent interruptions inherent in RES. However, power networks, like the current systems, must deal with problems related to rollout. The necessary changes to prevent this have been proposed since 2017 by energy regulators in various countries to identify and establish ES's playground. Significant changes include (1) the legal definition of "electricity storage" in ES, (2) balancing charges and the elimination of dual network, (3) identification of existing reserves in support of levy policies with the help of renewable sites, (4) operating network owners of reserves, (5) defining simplified planning clearance for ES, and (6) identifying de-rating factors used for storage in the capacity market. Because the regulation proposals for change at the levels of different countries are identical, Asia has been identified as a case study as part of its corresponding regulations and their market design. Based on the SWOT analysis, current regulatory changes in Asia examine whether BESS's business capability and business case are relevant. The following benefits have been identified for ESs due to these changes. First of all, reducing the cost for ES maintenance and charging for a double network means that the charges can be reduced by eliminating the demand residue. The government will then be able to obtain subsidies with the RES sites for all compatible resources for the ES to consolidate a better business case for the colocation of the ES. Finally, it was necessary to design infrastructure and cost-effective access for a positive impact on ES investment decision-making and easier access to the grid.

The current paper discusses BESS as a small-scale system of power systems for ES, which plays a key role. Storage properties should be created independently under the supervision of regulators without interfering with the current market construction requirement. The power of the traditional generator of ESs in favor of technological advancement is being researched to improve its density. Much of the current case involves long-term regulation of market regulation and energy sectors that limit the value of BESS. For example, short-term payments are made in CMs with a longer-term system.

Multiple opportunities have been found for private parties to obtain ES equipment by overcoming previous shortcomings for higher profits and by generating multiple revenues through a variety of services while bearing the high capital cost of BESS, as well as promoting and owning their own. Private parties are also at the forefront of approving and providing several integrated units for ESs in the wholesale market by the recently prescribed regulations. In addition to the analysis of the results of the SWOT method, the major problems caused by them are also mentioned in this paper. The first of those issues is the need to quickly generate independent power markets for ES based on power reserves in the current era, and the current definition must reflect the current characteristics of ES. Finally, the potential for BESS, which is a de-rated asset in the CM, is affected by the decline, and an economic assessment of the desired cost, as well as the power of the ES power network to be implemented.

References

BEIS. (2017). *Contracts for di. Contracts For difference government response to the consultation on changes to the CFD contract and CFD regulations* (n.d.) London, UK: UK Government.

Bradbury, S., Hayling, J., Papadopoulos, P., & Heyward, N. (2015). Smarter network storage electricity storage in GB: SNS 4.7 Recommendations for Regulatory and Legal Framework (SDRC 9.5). Available online: http://innovation.ukpowernetworks.co.uk/innovation/en/Projects/tier-2-projects/Smarter-Network-Storage-(SNS)/ (accessed on 16 April 2019).

Bradbury, S., Laguna, A., & Papadopoulos, P. (2016). *Electricity storage in GB: Final evaluation of the smarter network storage solution (SDRC 9.8).*

Brouwer, A., van den Broek, M., Seebregts, A., & Faaiji, A. (2015). Operational flexibility and economics of power plants in future low-carbon power systems. *Applied Energy, 156*, 107–128.

Castagneto Gissey, G., Dodds, P., & Radcliffe, J. (2018). Market and regulatory barriers to electrical energy storage. *Renewable & Sustainable Energy Reviews, 82*, 781–790.

Crotogino, F., Mohmeyer, K., & Scharf, R. (2001). *Huntorf CAES: More than 20 years of successful operation*, [Online]. Available: http://www.uni-saarland.de/fak7/fze/AKE_Archiv/AKE2003H/AKE2003H_Vortraege/AKE2003H03c_Crotogino_ea_untorfCAES_CompressedAirEnergyStorage.pdf.

Ding, J., Xu, Y., Chen, H., Sun, W., Hu, S., & Sun, S. (2019). Value and economic estimation model for grid-scale energy storage in monopoly power markets. *Applied Energy*, *240*, 986–1002.

Dinorwig Power Station. (1984). [Online]. Available: http://www.fhc.co.uk/dinorwig.htm.

Dominion, Bath County Hydro Storage. (1985). [Online]. Available: http://www.dom.com/about/stations/hydro/bath-county-pumped-storage-station.jsp.

Eller, A., & Gauntlett, D. (2017). Energy storage trends and opportunities in emerging markets. Navigant Consulting Inc.: Boulder, CO, USA.

Faunce, T., Prest, J., Su, X., Hearne, S., & Iacopi, F. (2018). On-grid batteries for large-scale energy storage: Challenges and opportunities for policy and technology. *MRS Energy Sustain*, *5*, E11.

Fulbright, N. (2022). *Regulatory progress for energy storage in Europe*. Availabe online: https://www.nortonrosefulbright.com/en/knowledge/publications/8b5285f4/regulatory-progress-forenergy-storage-in-europe (Accessed on 18.04.19).

Gailani, A., Al-Greer, M., Short, M., & Crosbie, T. (2020). Degradation cost analysis of Li-Ion batteries in the capacity market with different degradation models. *Electronics*, *9*, 90.

Gailani, A., Crosbie, T., Al-Greer, M., Short, M., & Dawood, N. (2020). On the role of regulatory policy on the business case for energy storage in both EU and UK energy systems: Barriers and enablers. *Energies*, *13*(5), 1080.

Gamble, F., Osiecki, J., Cais, M., Pishardy, R., Disalvo, F., & Geballe, T. (1971). 493–497). Intercalation complexes of lewis bases and layered sulfides: A large class of new superconductors. *Science (New York, N.Y.)*, *174*.

Geske, J., Green, R., & Staffell, I. (2020). The cost of bilaterally uncoupling British-EU electricity trade. *Energy Economics*, *85*, 104599.

Graditi, G., Ippolito, M., Telaretti, E., & Zizzo, G. (2016). Technical and economical assessment of distributed. *Renewable & Sustainable Energy Reviews*, *57*, 515–523.

Hamelink, M., & Opdenakker, R. (2019). How business model innovation a storage market. *Renewable Energy*, *131*, 120–127, *2019*.

Hossain, M., Madlool, N., Rahim, N., Selvaraaj, J., Pandeyx, A., & Khan, A. (2016). Role of smart grid in renewable energy: An overview. *Renewable & Sustainable Energy Reviews*, *60*, 1168–1184.

Huang, J., Fan, J., & Furbo, S. (2019). Feasibility study on solar district heating in China. *Renewable & Sustainable Energy Reviews*, *108*, 53–64.

Hübler, M., & Löschel, A. (2013). The EU Decarbonisation Roadmap 2050—What way to walk? *Energy Policy*, *41*, 255–272.

IEA. (2017). World energy investment 2017. IEA.

IRENA. (2017). *Electricity storage and renewables: Costs and markets to 2030* (n.d.) *Abu Dhabi, UAE: International Renewable Energy Agency*.

Khan, A., Verzijlbergh, R., Sakinci, O., & De Vries, L. (2018). How do demand response and electrical energy storage affect a capacity market? *Applied Energy*, *214*, 39–62.

Kocer, C., Cengiz, C., Gezer, M., Gunes, D., & Cinar, A. (2019). Assessment of battery storage technologies for a turkish power network. *Sustainability*, *11*, 3669.

Kötz, R., & Carlen, M. (2000). Principles and applications of electrochemical capacitors. *Electrochimica Acta*, *45*(15–16), 2483–2498.

Lockwood, M., Froggatt, A., Wright, G., & Dutton, J. (2017). The implications of Brexit for the electricity sector in. *Energy Policy*, *110*, 137–143.

Mayer, P., Ball, S., Vögele, S., Kuckshinrichs, W., & Rübbelke, D. (2019). Analyzing brexit: Implications for the electricity system of Great Britain. *Energies*, *12*, 3212.

Narayanan, A., Mets, K., Strobbe, M., & Develder, C. (2019). Feasibility of 100% renewable energy-based electricity production for cities with storage and flexibility. *Renewable Energy*, *134*, 698–709.

NYPA. (2010). *Blenheim–Gilboa pumped storage power project*. Retrieved from http://www.nypa.gov/facilities/blengil.htm: [Online]. Available: http://www.nypa.gov/facilities/blengil.htm.

OFGEM. 2017. Clarifying the Regulatory Framework for Electricity Storage: Licensing; Office of Gas and Electricity Markets: London, UK, 2017. Available online: https://www.ofgem.gov.uk/publications-and-updates/clarifying-regulatory-framework-electricity-storage-licensing (accessed on 17 April 2019).

OFGEM. (2018). *Feed-in tariffs: Guidance for licensed electricity suppliers (Version 10)* (n.d.) London, UK: OFGEM.

Pearce, P., & Slade, R. (2018). Feed-in tariffs for solar microgeneration: Policy evaluation and capacity projections using a realistic agent-based model. *Energy Policy*, *116*, 95–111.

Peker, M., Kocaman, A., & Kara, B. (2018). Benefits of transmission switching and energy storage in power systems with high renewable energy penetration. *Applied Energy*, *228*, 182–1197.

PowerSouth, Compressed Air Energy Storage. (1991). [Online]. Available: http://www.powersouth.com/mcintosh_power_plant/compressed_air_energy. (n.d.).

Poyry. (n.d.). Storage business models in the GB market. *Pöyry Management Consulting: London, UK, 2014.*

Schermeyer, H., Vergara, C., & Fichtner, W. (2018). Renewable energy curtailment: A case study on today's and tomorrow's congestion management. *Energy Policy*, *112*, 427–436.

TVA. (1978). Raccoon valley pumped hydro facility [Online]. Available: http://www.tva.gov/sites/raccoonmt.htm.

Wang, Q., & Li, R. (2016). Impact of cheaper oil on economic system and climate change: A SWOT analysis. *Renewable & Sustainable Energy Reviews*, *54*, 925–931.

Whittingham, M. (1976). B The role of ternary phases in cathode reactions. *Journal of the Electrochemical Society*, *123*, 315–320.

Whittingham, M., & Huggins, R. (1972). BBeta aluminaVPrelude to a revolution in solid state electrochemistry. *NBS Special Publications*, *364*, 139–154.

Winfield, M., Shokrzadeh, S., & Jones, A. (2018). Energy policy regime change and advanced energy storage. *Energy Policy*, *115*, 572–583.

Yao, Y., & Kummer, J. (1967). BIon exchange properties of and rates of ionic diffusion in beta-alumina. *Inorganic Nuclear Chemistry*, *29*, 2453–2466.

Chapter 3

Smart home energy management system: concept, architecture, infrastructure, challenges, and energy management

P. Muralidhara Rao[1], Reddi Sivaranjani[2] and P. Saraswathi[3]
[1]School of Computer Science and Engineering, Vellore Institute of Technology, Vellore, India, [2]Department of Computer Science and Engineering, Anil Neerukonda Institute of Technology and Sciences, Visakhapatnam, India, [3]Department of Computer Science and Engineering, GITAM Institute of Technology, Visakhapatnam, India

3.1 Introduction

Of late, the smart home energy management system (SHEMS) has become ubiquitous in advanced technologies, and these systems' deployments have increased quickly. It has gained much popularity for the use of smart energy management in the public and private sectors. Necessary applications allow the SHEMS to accomplish various demanding services like network controlling, offloading, peak shaving, demand response, load shifting, and other supplementary services. It has been extensively used for advanced energy management solutions in smart homes. Moreover, numerous research works have been carried out to use energy management that dealt with energy consumption efficiently. Besides, research on SHEMS has been emerging across several disciplines, especially for smart home energy usage. The main objective of SHEMS is to ensure energy savings, minimized peak average ratio, low electricity consumption cost, peak demand and supply, and reduce energy wastage (Rathor & Saxena, 2020). The SHEMS has been an existing mechanism for the past decades. It has served numerous functionalities such as monitoring, controlling, and optimizing energy consumption and flow (Mahapatra & Nayyar, 2019). Besides, it has several difficult electrical network applications for power generation, transmission, and distribution

Sustainable Networks in Smart Grid. DOI: https://doi.org/10.1016/B978-0-323-85626-3.00005-3

services. It has various applications that signify energy management systems with supervisory control and data acquisition.

Many basic energy sources for electricity, on the other hand, are currently scarce and expensive. Coal, oil, and natural gas prices all increase every year. Solar and wind are limitless fundamental energy sources, but the technology to fully collect their energy content is still developing. On the other hand, worldwide electricity consumption continues to rise year after year (Armaroli & Balzani, 2007), mostly due to increases in global urbanization (Especially in India and China). Furthermore, because of dynamic expectations of residential customers in terms of convenience, comfort, and adaptability, annual home energy consumption has been steadily increasing. Nowadays, energy management has become a significant concern and an essential part of human lives. However, energy management in household, public, and private service sectors must be considered to reduce the environmental burden to enable socio-economic impact on future generations. In recent times, research interest towards the development of the Smart Grid (SG) has become ubiquitous (Balzani & Armaroli, 2010).

Smart Grid is a smart electricity network to administer and assign energy cost-effectively and sustainably in comparison to a standard network. However, the ability to utilize the SG, requires a home that is communicative and adaptable. According to the literature, a smart house should include four systems: (1) a building automation system, (2) a room automation system, (3) a telecommunication system, and (4) a computer-assisted facility management system. SHEMS is comprised of five sections. The measuring device is the first component, and SHEMS relies heavily on measuring instruments for control. The principal measurement instruments for households are gas, water, and electricity meters (Pan, Jain, & Paul, 2014). The introduction of Advanced Metering Infrastructure (AMI) (Mohassel, Fung, Mohammadi, & Raahemifar, 2014) allows for the measurement of detailed, time-based data and the gathering and transmission of data to specified destinations regularly. The sensing device is the second component. Third one belongs to current, voltage, temperature, motion, light, and occupancy detection are all household sensors relevant for SHEMS (Avancini et al., 2019).

Energy is a necessary component of life. Electrical energy has become a basic need for civilization and a critical element for socio-economic growth in the current world as built-in overall energy necessities. Electricity is a right of every resident in the Netherlands; however, it must be paid for according to market norms. It would be a game-changer if such types of energy were infinite, and users paid fixed yearly or monthly rates regardless of their usage. Only system limits and administrative procedures could limit energy supply for energy suppliers and grid operators. Signals are sent to a central authority based on the desired parameters at various places. Smart appliances could be administered (Moriarty & Honnery, 2010), scheduled, or controlled in order to run at specific times using these criteria. The third component is ICT enablement, sensor, meters, and devices linked to the monitoring or control unit via the ICT. For the integration of numerous residential gadgets, both wireless and cable communication technologies are being developed. Some prominent technologies for facilitating home area networks are Wi-Fi, Z-wave, Home-Plug, and networking (Sodiq et al., 2019). There is no true winner in an overview of the leading communication protocols for home area networks.

With the rapid growth of the population, environmental constraints, diversification of city bounds, and power consumption are rapidly increasing. The electricity grid must be transformed into a efficient modern, secure smart grid environment that incorporates two-way communication, information and communication technology (ICT), and intelligent procedures to achieve this. Demand-side management (DSM) is an agile strategy utilized by services to reduce their customers' energy costs through demand response (Dr) programs. Renewable energy resources (RER), such as tiny wind turbines and roof-top solar PV panels, have been placed at residential buildings to minimize conventional energy costs (Kuru & Ansell, 2020). Furthermore, due to the intermittent nature of RER, consumers utilize battery storage systems (BSS) to mitigate variations. The integration of new generation resources is expected to have an impact on dynamic pricing. With the widespread usage of smart residential components and IT integration, there has been a surge in technological improvement of residential structures to become smarter. Smart components must be correctly scheduled to get the most out of real-time pricing (RTP) and maximize RER utilization. The researchers have not yet assumed the degradation in battery cost in the demand-scheduling software to explore the influence on practical factors and operational costs. In the authors' optimized scheduling of storage devices, deterioration of battery cost was used as the objective function. The developers (Mohammadi et al., 2018) have outlined the depth of discharge of a battery for every half cycle in a secured pricing model for BS in electricity markets. The researchers (Yang, Li et al., 2020) have proposed a novel framework for calculating the battery's half or full-cycle compared to the rain-flow approach.

Modern SHEMS are naturally vulnerable to cyber assault, so cyber-attack-resistant scheduling strategies are required. SHEMS' current scenario could lead to more charging and discharging cycles, reducing battery life. As a result, demand scheduling formulas must account for the expense of battery degradation and user comfort. The current research aims to design a extensive scheduling problem for minimizing energy cost in view of battery degradation. This chapter deals with various SHEMS mechanisms, security and technological challenges associated with smart grid networks (Qi, 2019). It is important to note that the integration of battery degradation costs and remission of the effect of a cyber-attack in the organizing problem of SH design has not been done before. As a result, an attempt is made in this paper to reconstruct the goal formula by including FDIA limitations and battery degradation cost on the process of scheduling. This research aims to develop FDIA-resilient SHEMS scheduling that considers pricing and battery degradation costs in the context of objective function. The current study looks at an FDIA assault that uses the Gaussian process regression (GPR) approach and tedious tasking to identify commonly utilized detection tools (Yang, Zhao, Li, & Zomaya, 2020).

3.2 History and evolution of smart home energy management system

The name SHEMS stands for "smart home energy management system," which it is derived from "energy management system" (EMS). In the home sector, EMS has a long history of use. Workable night thermostats have been used as a kind of automatic energy control since the early 1900s. However, in the 1970s, energy management became a major concern as a result of recurrent energy crises, rising costs, and the concept of energy conservation. Honeywell created a one-of-a-kind solar EMS based on microprocessor technologies in the 1970s as a key contribution to resolving the energy issue. The EMS was significantly altered by technology advancements in the 1980s, especially with the introduction of PCs. Companies such as General Electric, Hitachi, Siemens, and Toshiba developed the EMS, which used proprietary hardware and operating systems (Usman & Shami, 2013).

In the late 90's, software platforms such as LINUX, UNIX (Mackenzie, 2005), and Windows-based systems expanded the range of energy management system solutions. To propose a functioning and user-friendly energy management system at the home stage necessitates a distinct strategy than the current EMS in transmission & distribution systems. Recent advancements in embedded systems technology have improved the functionality of EMS even more. Most of those bulky, space-consuming solar cells technologies have made impetus to embedded or chip-based systems that are more portable, small, and cheap. SHEMS, like any other energy management system, aims to save energy, reduce costs, and increase comfort. SHEMS essentially provides five major services: monitoring, recording, control, managing, and alerts.

3.2.1 Possible technologies and solutions

Different approaches including AMI (Ikpehai, Adebisi, & Rabie, 2016) smart sensor technologies (Ahmed, Kang, & Kim, 2015), home area network (HAN) (Collotta & Pau, 2015), smart home appliances (Bradac, Kaczmarczyk, & Fiedler, 2015), and home energy storage system (HESS) (Pascual, Sanchis, & Marroyo, 2014) have been researched in recent years. Using authoritative control and data acquisition with energy management systems, including the transmission, generation, and distribution systems of the given electrical network, specifically in the notion of the smart grid, is referred to as SHEM. This approach has gained widespread acceptance as a way to forecast the future evolution of power systems.

SHEM has turn into a critical element for household consumers, similar to the home system, in order to successfully assign demand-side management in view of smart grids (Pau, Collotta, Ruano, & Qin, 2017). EM is concerned with the practical monitoring and organizing several household electronics in smart households, users' priorities, through intelligent ambient systems monitored by a human–computer interface, with the goal of lowering power costs and increasing energy consumption throughput (Fadda, Fadda, Ghiani, & Pilloni, 2019) owing to huge worries about global energy efficiency and environmental pollution, SHEM has also been created to monitor residential renewable energies (wind turbines, solar panels, and so on). With the quick growth of renewable energy (RE) and advanced power electronics, few smart sensors can be integrated into SHEM in order to ensure the final usage of the preserved energy sources equipped in house-hold developments, resulting in improved indoor home energy conversion and utilization efficiency (Sepasgozar et al., 2020).

SHEM has piqued the interest of the research community in recent years as a result of these factors. The fundamental reason is that incorporating current automation technology into the home can result in significant energy savings while also lowering the building's operational costs over its entire lifecycle. However, because of the complexities of the existing systems, and also the common issue of in-sufficient monitory methods, smart residences' full potential remains untapped. As a result, energy usage is higher than it should be, and consumers are not able to achieve complete assurance in their digitized homes.

SHEMS solutions have progressed from pilots to fully operational programs in some cases. During initial tests, the PowerMatcher technology resulted in a 30% peak reduction in heat crisis if residency owners were willing to adopt a one-degree variance in temperature (Asare-Bediako, Ribeiro, & Kling, 2012). In buildings, DOMOSEC accompanied by autonomous floor monitoring modules is used to increase energy efficacy. AMIGO, ACHE, eDiana, SOCRADES, IRS and AIM are examples of smart technology. SHEMS are predicted to rise in the upcoming generations, especially for novel household structures, due to advancements in both wireless and wired technology, as well as the utilization of embedded system technologies.

3.3 Smart home energy management system: overview

Smart houses, for example, have benefited from the expansion of the Internet of Things (Talari et al., 2017). Today's smart home uses go beyond luxury, with EMS, for example, allowing for more efficient energy usage. Industrial IoT providers are typically focused on developing features, including the small devices with a wide range of communication and sensing capabilities. However, security and privacy have always been a big worry, preventing IoT devices from being widely adopted. The SHEMS is a technological framework that consists of software and hardware that permits a user to control energy consumption and generation, as well as man-monitored and digitize energy use inside a house." To account for the uncertainties of electricity price and load, Raja (Raja & Baskar, 2012) described a Dr application that uses the two-point estimate approach and particle swarm optimization while retaining an extreme level of accuracy. Ahmad et al. (2017) presented an optimized home energy management system (OHEMS) that schedules in response to the dynamic price of electricity, household appliances and ESS have been developed to lower the cost of electricity (Table 3.1).

Qureshi et al. (Xue, 2017) used SHEMS implemented on a Raspberry Pi 3 Model B to lower electricity bills by using user-defined priority-based load scheduling in response to real-time power pricing. Through data sharing between HEMS and grid energy management system, Yang et al. (2017) carried out a study that used PEVs charge-discharge control to optimize PV output utilization and reduce PV power forecast error, cutting the system's operation costs. Shareef, Ahmed, Mohamed, and Al Hassan (2018) offered a HEMS case study for a home customer that wanted to participate in Dr with smart plugs. Customers' comfort levels are considered when Dr is implemented utilizing an appliance load monitoring mechanism. Finally, Misleh (2014) carried out a SHEMS-based stochastic framework. PV, water heaters, and batteries are incorporated in SHEMS modeling, where energy acquisition costs are in the wholesale electricity market, costs are minimized.

The SHEMS has two other controls. To begin, homeostatic control (HC) was first proposed by Schweppe et al. in the early 1980s and is based on the regulation of homeostasis in a living organism. To maintain a steady internal

TABLE 3.1 features of smart home energy management system (SHEMS) and BEMS.

BEMS	SHEMS
It has dynamic demand control and response	It optimizes home energy usage
It minimizes energy consumption within the home	Decision is relied on tariff and pricing policy
It defines short-term forecast of building load	It controls effective Dr participation
It minimizes waste of thermal/ heat energy	It has consumer involvement and interaction in energy management and conservation

equilibrium, organizations constantly alter their intake and energy expenditures by interacting with their surroundings. Likewise, by communicating with clients, HC seeks to maintain a balance between demand and supply. As a result, domestic customers are constantly informed of changing electricity rates and have the freedom to choose their consumption rather than being forced to follow the prescribed electricity consumption. The microgrid connected to the utility grid to give power to a group of residential users is where HC finds its most widespread application. Microgrids can communicate with utilities and customers in real time and adapt to changing conditions to maintain supply and demand balance that benefit both the customer and utility (Yamashita, Vechiu, & Gaubert, 2020).

3.3.1 Components

The SHEMS is well known for developing a smart home system that monitors and manages electrical appliances, including load, storage, or generating resources. Monitoring, logging, control, administration, and alarm services are examples of SHEMS functional modules. Zhou et al. (2016) integrated three separate viewpoints: functional, physical, and communication, to create a smart house reference architecture depicted in Fig. 3.1. The SHEMS can be determined based on five categories (Saad al-sumaiti, Ahmed, & Salama, 2014). The measurement device is the first component; it relies heavily on measuring instruments. The principal measurement instruments for households are gas, water, and electricity meters (Schnieders & Hermelink, 2006). The introduction of AMI allows for the measurement of detailed, time-based data and the gathering and transmission of data to specified destinations regularly. The vital components of smart home systems are described below, based on this design.

1. *IoT (Internet of Things) devices*: The major roles of these are to perceive the environment, receive commands and communicate data. As a result, they have communication abilities and may communicate with another HAN parts like IoT hubs, home gateways, or other IoT related communicating devices. Metering devices and controlled loads are examples of IoT devices in SHEMS.

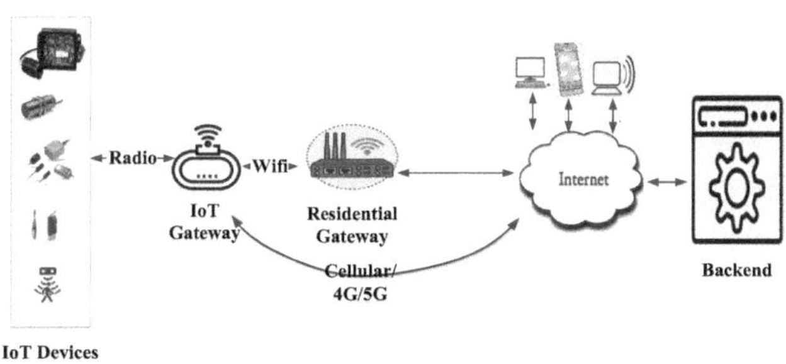

FIGURE 3.1 A typical Smart home system architecture.

2. *Gateway:* It links HANs and Wide Area Networks to connect IoT devices to the Internet (Abdulrahman et al., 2021). (WAN). A gateway can operate as an IoT hub or vice versa in some situations.
3. *Backend:* It is a remote, centralized component that regulates numerous smart houses. It uses the Internet to communicate with the IoT hub and performs storage, monitoring, and control tasks for IoT devices. In addition, backend systems use APIs to connect to external programs, allowing connection with SHEMS (Shrestha, Johansen, & Noll, 2019).
4. *Hub for IoT*: It serves as both a central controller and a link between IoT devices and the backend system. The IoT hub receives sensor data and converts it before sending it to the backend system. Likewise, the IoT hub may receive control commands that it can pass on to the appropriate devices. The IoT hub, as opposed to IoT devices, has significantly more computing power and can make decisions about how to manage and control IoT devices (Ammar, Russello, & Crispo, 2018).
5. *Data from the application and the network:* The network data consists primarily of information about connectivity, whereas application data consists of meter values, commands for controlling devices, log data, and software image representations, among other things. IoT devices generate metering values, which are then delivered to the data and subsequently sent to the IoT hub, which ultimately sends it to the backend systems for storage and analysis. The IoT hub, on the other hand, receives control commands from the databases and sends them to the IoT communicating devices for execution (Krishna, Rao, & Rao, 2021; Rao & Saraswathi, 2021).

3.3.2 Infrastructure

A Building Energy Control System is a collection of energy sources and monitoring equipment that helps in the computer-assisted management of a building's system. Increased generation and lower operational costs need the use of BEMS. Buildings with feedback devices, as well as buildings with a feedback and feed forward method, are managed. Smart buildings are made up of smart devices. Both of these systems should be included in a SHEM. There are five basic components to the SHEMS.

Communication system is a channel for connecting and sharing information between items. The communication system is a critical component in SHEMS, and it is largely reliant on it. When we want it to be breach proof as well as open source, choosing a dependable and low-power communication solution is crucial.

1. Networks: Wireless networks are becoming increasingly prevalent and are now a part of our daily lives (Mohassel et al., 2014). Mobile network technology (GSM, GPRS), Bluetooth low energy, IEEE 802.11 (Wi-Fi),

and IEEE 802.15 are all included (ZigBee). Wi-Fi and ZigBee are popular choices because of their range and ability to handle many users. However, ZigBee does not have the necessary support to connect with our everyday gadgets. As a result, it necessitates the use of additional devices and software to connect, making it more complicated. Wi-Fi and Bluetooth low energy are more promising because they have no connectivity concerns and use less power (Mahmood, Javaid, & Razzaq, 2015).

2. PLC stands for Power Line Communication, and it refers to the use of an existing electrical system for communication purposes. PLC applications in SHEMS are still in their infancy. When wireless networks are unavailable, it is employed (Tekbıyık Ersoy, 2012) discusses PLC-related difficulties and potential challenges.

System of central management is composed of a central controller that oversees energy management as well as data collection from sensors and smart meters. It examines the data collected using code that includes load balancing and optimization algorithms. The results are then displayed on the HMI, along with a suggested solution on the GUI. The CMS can handle critical circumstances on its own. It also provides consumers with monthly and daily statistics on energy generation and consumption.

Smart meters and sensors are the primary devices that interact with the environment and collect relevant physical data. Smart sensors are devices that detect and transmit signals for feedback and are linked to a communication system. The majority of smart sensors have numerous functionalities. Smart Home requires a network of simple and smart sensors to function successfully. The programmable electronic devices known as smart meters may measure, control, and display system inputs. Measurement, control, communication, power management, and display functionalities are all included (Ghasempour, 2019).

Demand management is another important step to set up an EMS, it is necessary to have information about the load of a Smart Home. The loading of a SHEMS fluctuates over time and according to the consumer's needs. Improper loading can result in overloading, waste, and energy loss. Energy conservation and energy efficiency are two approaches to managing a SHEMS' demand. Energy conservation includes managing the behavior of the people who live in the Smart Home. It necessitates the precise application of electricity. This sort of management function may also be addressed by the smart nature of CMS, and this management returns in another form known as Demand Side Management, which involves disregarding the CMS's proposed savings. By analyzing the consumer's power usage and energy pricing, DSM alerts them to utilize alternative loads at a recommended time indicated by the Central Management System. Demand response and load shifting are the two primary DSM methods (Mohammadi, Faghihi, Kazemi, & Salemi, 2021). The usage of energy-efficient items or appliances is the second method of management. The use of smart lighting

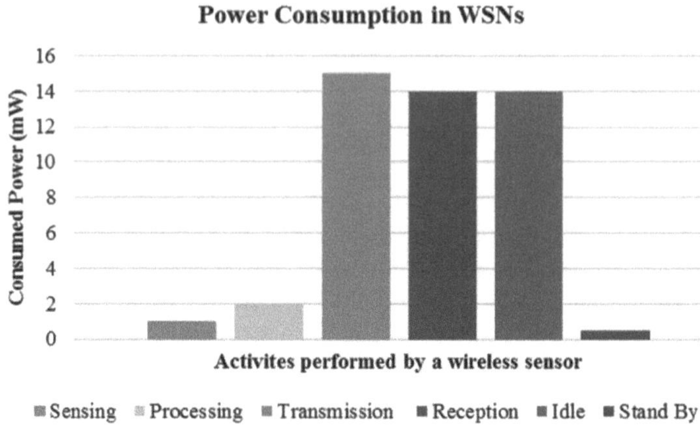

FIGURE 3.2 Energy consumption of various modules in Wireless Sensor Networks.

systems and daylight harvesting is becoming increasingly prevalent. It is suggested that 100-W bulbs be replaced with 25-W energy savers. This has the added benefit of conserving energy as well as money. These energy savers are also available in low-power DC versions, which are more dependable. Low voltage intelligent lighting systems with dimming solutions are available in both AC and DC versions and are 87% more energy efficient (Stoyke, 2006). There are also energy branded fans on the market that use less electricity. Because of the electricity scarcity and frequent disruptions in the utility power supply, integrating RE and energy storage devices is a feasible and significant energy management solution in Pakistan. It decreases the strain on the utility side while also saving money and energy.

Supervision of the power supply management aims for increased output and lower energy costs, RE generation at Smart Homes requires monitoring and effective control. The integration of many energy sources adds to the complexity of the EM system. Because RE is inherently intermittent, effective energy balancing strategies between demand and generation are required. It aids in the stability and long-term viability of the Smart Home electricity system (Olatomiwa, Mekhilef, Ismail, & Moghavvemi, 2016). The consumer may mix RE, utility electricity, and generator backup with smart energy power supply. The SEPS system is then linked to CMS (Fig. 3.2).

3.4 Energy management

Solar PV: power generation is influenced by total solar irradiation, weather conditions, and ambient temperature. The solar power, P_s, is calculated using the subsequent formula.

$$P_s = \mathrm{DF} \times P_{\mathrm{STC}} \times \frac{G_A}{G_{\mathrm{STC}}} \times [1 + (T_C - T_{\mathrm{STC}}) \times C_T \tag{3.1}$$

PSTC stands for nominal PV array power in kW under standard test conditions (STC), GA is for global solar radiation in kW/m^2, and TC stands for solar PV cell temperature, where DF is the derating factor, which is usually 0.8. The temperature under STC, which is equivalent to 25°C, is denoted as TSTC. CT can be defined as the PV temperature coefficient of −0.0011 C. The following expression can be used to compute TC from ambient temperature (Ta) and global sun radiation G on the horizontal plane. NT stands for normal operating cell temperature, which is usually 48°C.

$$\mathrm{TC} = Ta + \frac{\mathrm{NT} - 20}{0.8} \times G \tag{3.2}$$

Wind turbine: The amount of power created by a tiny roof-top wind turbine in a residential building is determined by various factors, including the swept area, wind speed, and air density. A tiny wind turbine's extracted power (PW) in kW is calculated in the subsequent equation:

$$P_W = 0.5 \times C_P \times \rho + A \times V^3 \tag{3.3}$$

3.4.1 Energy consumption scheduling

The SHEMS specification strives to identify needed energy-saving features that are both cost-effective and convenient for home users. It refers to a set of devices and services for the industry, but it is not intended to be viewed as a single platform or product. The SHEMS specification, on the other hand, is designed to allow service providers to test business models inside a standard framework while working with the ENERGY STAR program. As part of the certification process, smart home device makers are urged to collaborate with service providers (Mendes, Godina, Rodrigues, Matias, & Catalão, 2015). The following are SHEMS' overarching objectives:

- To ensure that gadgets are only used when needed (the seamless optimization of energy use, storage, and production while balancing cost, environmental impact, and convenience).
- To deliver an exceptional customer experience.
- Depending on industry standards and best practices, we can promote interoperability and security.

Some industry companies are also paying attention to SHEMS and advanced lighting systems, a segment that is undergoing significant transformation. For example, SHEMS plans to enhance smart water heater me-of-use pricing and incorporates grid services like demand response as it develops. According to Terry McGowan, the American Lighting Association's director of engineering and technology, these new systems will "give homeowners a quite different way to manage their lighting. Thus they can enjoy the best that lighting has to offer to light is an important aspect, from a utility perspective, from a consumer's viewpoint, and our industry's perspective."

The SHEMS specification should aid in the growth of the smart home market in the United States, but there are roadblocks. First, it is more difficult to combine hardware and services and define standards than just rate appliances or devices, which ENERGY STAR has accomplished admirably. Second, it will most likely be difficult to advertise SHEMS to customers effectively. Even though many people have purchased smart home products, millions more have not, and they are suspicious of their worth and concerned about data privacy risks. Third, SHEMS does not have an energy-savings measure at the outset, which must be added as soon as practicable.

3.4.2 Modeling and simulation

A MATLAB model is being developed to look into several areas of SHEMS, including peak load management, control algorithms, and smart grids, and predictive maintenance and scheduling, with a focus on the extra value SHEMS may provide to household customers.

The process of optimizing energy consumption is demonstrated in this study. The following are basic components of SHEMS' architecture for analyzing optimization, as shown in Fig. 3.3:

1. *Residential loads*: Independent (un) controllable loads and aggregated loads are the two types of residential loads. These loads are modeled using data collected from household customers or Simulink simpowersystems simulations.
2. *Onsite generation*: Photovoltaic systems, micro-CHps, or solar collectors, are present in some residences for inside propagation. The observed and simulated generations are used in the model. A heat pump is classified as a "prosumer" since it serves as both a heat generator and an electrical load.

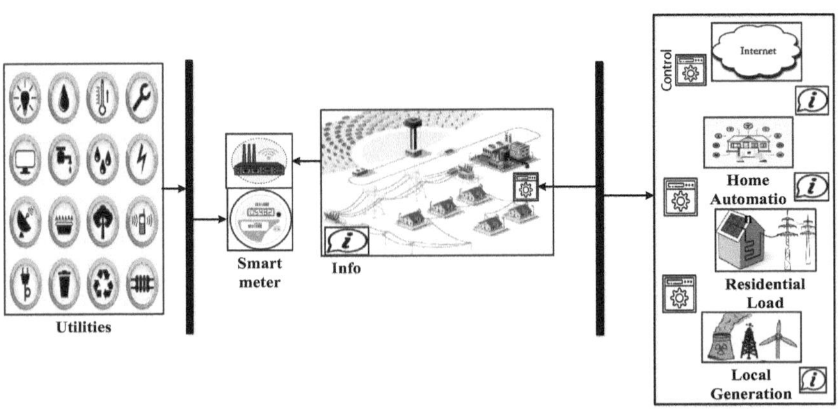

FIGURE 3.3 Smart home energy management system architecture for optimization analyses.

3. *Home automation*: Sensors and alarms are part of home automation. This module is designed as an empty module for the purposes of this research.
4. *Utility*: The energy block is used to deliver signals to households depending on the present network state. This is accomplished through the smart meter installed at the customer's location.
5. *Smart meter*: In the Netherlands, the smart meter is defined as the connection amidst some of the utilities and the residential. The signals generated from the services are sent to the users' SHEMS via the PI connection on the smart meter for further process of automation. It even calculates rates by measuring energy consumption and ne put into the grid independently.
6. *Internet*: For the SHEMS, the internet module is designed as a data communication center. This block retrieves variable tariffs, weather data, and other important information. It is the location where the customer's data is shared with third parties.
7. *Energy management system (EMS)*: EMS acts as the heart of intelligence and control. This block processes all of the data that has been measured, sensed, downloaded, and transferred. By balancing generation of demand or lowering costs, relying on the situation, EMS also monitors energy usage.

3.4.3 Home energy management

The structure of home automation is a technological framework that has been around for a while. The employment of thermostats to maintain heat levels in the early 1900s was a significant step toward assuring energy efficiency. Security systems, entertainment, energy-efficient appliances, HVAC control, ehealth, and remote controls have all been added to the system over time. The idea of SGs, with their underlying ICT, improves the effectiveness and efficiency of energy conservation in the power sector. Though the growth of ICT systems far outpaces that of other sectors in terms of energy consumption, it can help other industries save energy by using sophisticated monitoring and management systems (Fig. 3.4).

SHEMS, as a micro-SG, can help reduce household appliance energy use by applying sophisticated administering and control technologies. The environment is a fantastic scaled-down version of the public-grid. It can also be called a "home micro-grid" because it includes distribution, generation, and load at one place.

"Energy optimization" refers to energy use in a way that is both effective and efficient. It shows up in lower bills or a desirable balance between local generation and usage for residential clients. Generally, household with onsite generating are comparatively preferable than average household inhabitants to adopt SHEMS. In this part, optimization strategies for a grid-connected household with on-site generating are studied. There is no local storage and

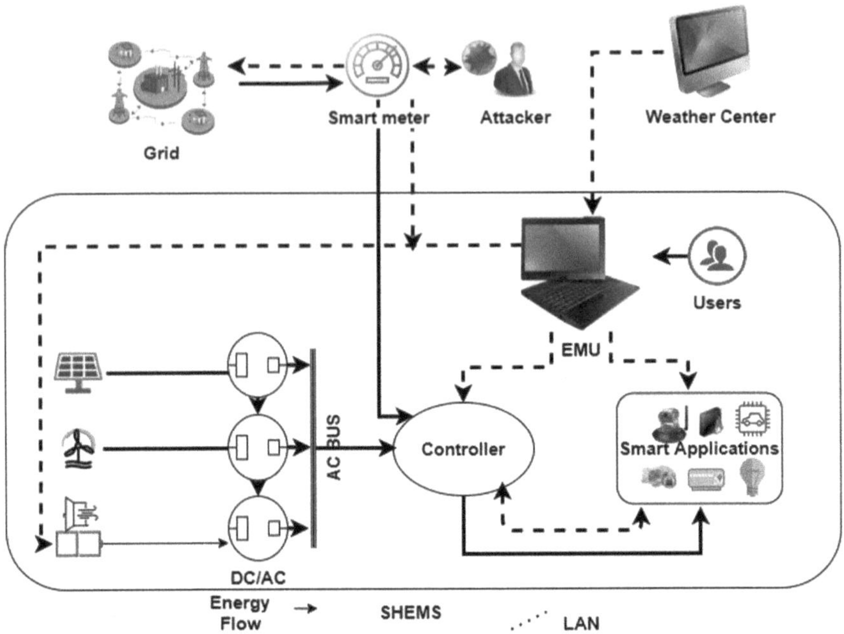

FIGURE 3.4 A generic smart home energy management system architecture.

energy flows in both directions (to and from the electric grid). For the integrated optimization, two parameters are chosen: the model looks into the possibility of SHEMS lowering expenses and increasing energy efficiency in a typical household. Variable packages are utilized for this process since they present a future dynamic tariff environment.

The proxy energy tariff for the Netherlands (Vasileiadou, Huijben, & Raven, 2016) has been considered for the evaluation, which is even utilized in the optimization process. Generally, residential consumers pay high amount of taxes per kWh of power consumed than they do for the real electricity used. Therefore, supply-demand optimization is often used to meet local consumption with onsite production. The model uses the following to carry out the optimization process:

1. Data is collected from a household client with solar panels, a heat pump, and solar collector, and a heat pump.
2. Living room, kitchen, bedroom, office, lobby, garage, and toilets are all rated as a group.
3. Each of the above appliances is estimated separately, with four of them being presumed to be adjustable loads. Among them are a dishwasher, a tumble dryer, a washing machine, and a heat pump (limited feasibility).

The various forms of organizing methods for decreasing energy expenditure in model SH have gotten a lot of attention in the literature. A particle swarm optimization technique was developed to reduce energy costs while maintaining consumer satisfaction in residential structures. The energy cost in a residential residence is reduced using a single goal multivariable optimization technique based on a genetic algorithm (GA). To reduce energy costs while also improving the peak-to-average ratio (PAR), multi-objective MINLP problem is constructed.

The PAR and energy cost of SH were lowered using a hybrid BAT pollination optimization technique. The energy cost and PAR are reduced via a dynamic pricing approach. Researchers presented a smart building using the dynamic Dr technique to identify the lowest energy price as function of objective. An artificial intelligence-based automated SHEMS is described for optimizing the operational discharge of DG units in household applications. Based on the consumer's spending goals, power consumption bill in SH was optimized. The authors suggested an energy-efficient scheduling algorithm for residential buildings that considered uncertainties in appliance operation times, the intermittent behavior of RE propagation, and BSSs. The HEM schedule mechanism can be used to cut electricity bills in view of a wide range of appliances and combining them into a united model. The authors used a inclining block rate model and combined RTP to reduce PAR and electricity costs.

The operational cost of SHEMS has been reduced by considering stochastic load behavior and consumption uncertainty. The authors presented a priority-based and optimization-relied dynamic Dr method to schedule demand of a smart home to lower the electricity expenditure. In HEMS, Tavakoli et al. (2018) suggested a conditional value at risk (CVaR) optimization strategy to mitigate the risk of practical energy price presentation and RER unpredictability. They employed a two-point estimating model (2PE) with CVaR to approximate solar power. To represent the uncertainties in HEM, the authors utilized a stochastic model. HEM is used to explain the uncertainties of electric vehicles that are now available, while RER is used to improve customer happiness by incorporating a fatigue response index. HEMS uses a multi-constrained mixed-integer problem (MCMIP) to reduce energy costs. The authors suggest a unique HEMS for practical electricity scheduling in household buildings to reduce energy costs while considering human interaction aspects, power supply unreliability, consumer preferences and priorities. The authors investigated residential Dr by scheduling household appliances to save energy costs, which resulted in a mixed-integer nonlinear programming issue (MINLP).

3.5 Technological issues and challenges

Smart homes (SHs) are vulnerable to cyber-attacks since they are incorporated with ICT and enhanced methods of intelligence. A user or a SHEMS provider could be the attacker. The hacker is thought to have injected bogus data into the smart meter in order to cause an error based on real and expected data for a particular range. The attacker merely injects fake data in this situation. Despite its apparent advantages, SHEMS is not without its drawbacks. Customers have reservations about SHEMS' effectiveness. Second, despite its growing market share, smart home solutions remain rather pricy.

Problems with information and education are also major roadblocks to SHEMS implementation. Residential clients are sometimes unaware of or misled about SHEMS' capabilities. Some see it as a waste of money, while others consider privacy and network security, as well as the health risks associated with expanding wireless networks at home.

3.5.1 Smart home energy management system implementation challenges

The process of implementing SHEMS goals presents various problems, the most significant of which is sustaining constant standards for SH&E quality in the entire Project. Project managing strategy loses accountability when it comes to SH&E predictions if they are inconsistent. Every member of project team should be dedicated to displaying the project's vision on reaching "World Class Safety Standards" on a personal level. Many of the Amherstia Project's work locations are on Contractor-monitored sites, which present few particular issues. These are regions where the project-contractor is in charge of daily operations and work is carried out in accordance with the Contractor's safety program. The Project-team is working by the help of Contractor to enhance the Contractor's safety program and implement required measures when the program does not meet project criteria.

Another problem is the current slump in the oil business, which puts immense burden on contractors to cut prices. Several contractors regard defense only as an expense, oblivious to the numerous advantages that come with great safety performance. For the contractor, safety is a capital that offers benefits in a variety of ways. The pivoting advantage to be noted is it lowers the expenses of injuries and diseases. Compensation premiums and medical fees have large direct costs. Furthermore, the built-in indirect costs, such as lost productivity, replacement labor, equipment damage, and time-management, have been demonstrated to be many times higher in the past.

Collaboration among industrial companies operating in the same fabrication fields has verified to have a significant impact on project-contractors. Furthermore, since one safety procedure will fulfill the needs of every customer, a standard set of predictions among industrial businesses optimizes the Contractors' efforts. The Contractor will be able to make enhancements to one's safety procedures that will be useful and have adequate worth for them in the long period if industry organizations have a consistent set of expectations.

Maintaining a commitment to keeping the SHEMS documentation current is one of the document's challenges. As the project teams moves in all stages of the project, the SHEMS must be updated on a regular basis. An obvious transition strategy from the construction team to operations team, for example, must be devised then implemented into the SHEMS. During this transformation, the team has also been insisted to strengthen the dedication to "World Class Safety." It will be easier to meet this issue if the ultimate leaders of the operations and drilling teams are on the Project and are already indulged with SHEMS process of implementation.

The SHEMS document must be audited on a regular basis to ensure that it follows the Contractor's safety policy as well as the predictions of other companies working on the sites of fabrication. Reviewing the SHEMS model could also be a good stab to renew the SH&E program with these outside stakeholders as well as among the project team.

The challenge for upcoming investment projects is to maintain contractors and BP Amoco capital project teams to the same high level of SHEMS requirements. This means that on all future projects, management predictions about "World Class Safety Standards" should persist the greatest choice. It will also be critical to disseminate the SHEMS standards used by the Amherstia Project to other BP Amoco project teams, as well as to discuss accomplishments and problems. Upcoming investment project teams will have commitment to address SHEMS concerns much sooner in the project deadline and keep efforts to enhance SHEMS management on an ongoing basis (Vandermolen & Cella, 1999).

3.6 Real-time solutions and recommendations

Controlling a building's lighting, security, comfort, and also other technology frameworks is a requirement that predates housing. There exists an immediate requirement to monitor and assign constraints that primarily influence heating systems in major corporate buildings. Since the 1970s, infrastructures with management modules for the regulation and monitoring of a variety of systems have been available, including electrical distribution, elevators, lighting, and, in particular, access control and anti-theft protection. New urban contexts have emerged in which man is present on a regular, continuous, or occasional basis, necessitating administration and control of structures other than corporate buildings. Smart home models can be implemented in the aforementioned scenarios using the building automation decalogue provided in (Chinchero & Alonso, 2020) Corporate structures, government structures, offices, industrial sectors, housing blocks, hospitals, universities, commercial and entertainment structures, and so on. In smart buildings, control networks are used to achieve welfare and comfort, as well as safety, energy efficiency, and communications (Chinchero & Alonso, 2020; Lorente, 2005).

3.6.1 Optimal analysis of energy management system using multi-objective method

The model assesses the levels of man ease-stay in houses as well as energy usage will be important aspects in civilized structures. The approach is based on human beings' constant state of contentment. It considers human comfort and energy savings independently, with the goal of balancing both using the Pareto model to resolve a multi-objective problem. The suggested intelligent building multi-agent control system even anticipates advancements in communication technologies and artificial intelligence (Antoniadou & Papadopoulos, 2017).

In general, networks of control have system incorporation and proposed tools that can be used to work on automation for buildings, home, transportation, industry, and other applications. For smart applications, a variety of network control protocols have been proposed, including Building Automation and Control Network (BACnet) KNX Standard, BUSing protocol, Local Operative Network (Lonworks), and others. Control Network solutions can make a significant contribution to the design and improvement of Smart Building methodology and control strategies (Shahzad et al., 2020).

For SB applications, centralized and distributed systems are used to implement automation and control technologies (Chinchero & Alonso, 2020). A framework may use distributed control, centralized control or a integration of both. In arriving at a centralized model, there exists a control module that oversees entire implementation and is connected to all calculations and performance aspects, like controllers and user interfaces (Chinchero & Alonso, 2020).

3.6.2 Smart home energy management system with dynamic pricing

Sensor automation and learning algorithms are employed in this situation, allowing SHEMS to effortlessly adjust the cost of residential usage practically. Water heating, heating, ventilation, and air conditioning, as well as electric vehicles, may all be controlled. In order to facilitate load control, the user can also engage with the service providers. The simulation is run on a prototype that models a simple Bayesian classifier using Bayes' theorem and a valid independent hypothesis (Jacoby, 2018). Temperature sensors, timing, illumination, real-time price, user operations, and other inputs are all considered in the suggested approach. This data is processed using artificial intelligence and deployment of algorithms for lighting, air conditioning, and load control, resulting in output vectors such as temperature control, turn on/off, activation and deactivation appliances, data recording, and so on.

3.6.3 Energy use optimization

It is another real time solution framework is provided to optimize the utilization of building infrastructures while preserving the users' basic standards of comfort. The results are shown using the Simscape software for simulation. The framework employs the Scheduler approach, which defines three states: supply monitoring, prediction, and real-time demand control. This type is likewise multi-agent, such as H-agent, which controls and regulates the room heating system. The ventilation system is monitored and controlled by the V-agent (room level). The ventilation system is monitored and controlled by the V-agent (room level). The C-agent is in charge of monitoring and controlling the cooling model (room level). E-Agent regulates and also monitors lighting and electrical systems at the room. Then U-Agent defines the load factors / weight agent's region and monitors and controls the occupancy level. It is also in charge of training the choice of customers, creating identities for them, identifying them, and providing feedback to the system (Hurtado, Nguyen, Kling, & Zeiler, 2013)

3.6.4 Multi-agent systems for smart home energy management system

In this example, a SHEMS framework is introduced. The authors in (Asare-Bediako, Kling, & Ribeiro, 2013) developed an agent-based system that implements procedures capable of handling complicated and dynamic decision distributed intelligence problems. The study presents a multi-agent architecture for efficient HEM, which can be used to evaluate future models for it. Cost, comfort, and green (energy-efficient) are the actors in this method for the model of smart demand management system. Then simulation is executed with curves that indicate quantity of energy spent by all devices in a home vs the energy cost, and the resultant data is fed into the Management and Optimization Agent (MOA) (Qi, 2019). MOA is in charge of determining an equilibrium pricing with the least amount of energy use. In this situation, software frameworks such as JAVA/JADE and MATLAB are used to run the simulation.

3.7 Future smart home energy management system

A concept for future SHEMS is outlined briefly. They show what factors should be considered while constructing long-term SHEMS and how they interact. It takes into consideration the building itself, such as the type of building, the year it was built, and the direction when constructing future SHEMS. The type of people who live in that house, such as the elderly, huge families, students, or singles, is also relevant. Second, the SHEMS' connection to the grid is critical for its integration and long-term viability. Residential demand response applications are possible with it. The third option is local installation at the customer's location. Do smart meters, sensors, and appliances exist, and if so, how are they integrated? Is there a distribution generator at the customer's location, and if so, how can it be integrated into the system? The fourth principle to be addressed is energy management.

Different appliances and solutions from other vendors must be supported by the ICT, system software, and algorithms discussed in (Stoyke, 2006). If the SHEMS solution is to be sustainable, regulatory, social-economic, and environmental considerations that are often disregarded throughout the design stages should be considered. Finally, it is critical to identify the SHEMS' primary stakeholders. This may differ from one location to the next, as well as from one country to the next. The client is at the center of it all, making the decision. The unpredictability of residential behavior is an important element to consider.

3.8 Conclusion

Nowadays, energy management has become a significant concern and essential part of human lives. However, energy management in household, public, and private service sectors must be considered to reduce the environmental burden to enable socio-economic impact on future generations. This chapter presents a systematic literature review on SHEMS, roles and responsibilities of various components, and benefits of SHEMS driving energy efficiency. In addition, we emphasize major concepts, components, functional capabilities, architectures, infrastructures, and existing energy management solutions for SHEMS. Lastly, we provide a comprehensive review of various energy management solutions, technological aspects, issues and challenges, and recommendations to enhance the performance of SHEMS.

References

Abdulrahman, L. M., Zeebaree, S. R., Kak, S. F., Sadeeq, M. A., Adel, A. Z., Salim, B. W., ... Sharif, K. H. (2021). A state of art for smart gateways issues and modification. *Asian Journal of Research in Computer Science*, 1–13.

Ahmad, A., Khan, A., Javaid, N., Hussain, H. M., Abdul, W., Almogren, A., ... Azim Niaz, I. (2017). An optimized home energy management system with integrated renewable energy and storage resources. *Energies*, *10*(4), 549.

Ahmed, M. A., Kang, Y. C., & Kim, Y. C. (2015). Communication network architectures for smart-house with renewable energy resources. *Energies*, *8*(8), 8716–8735.

Ammar, M., Russello, G., & Crispo, B. (2018). Internet of things: A survey on the security of IoT frameworks. *Journal of Information Security and Applications*, *38*, 8–27.

Antoniadou, P., & Papadopoulos, A. M. (2017). Occupants' thermal comfort: State of the art and the prospects of personalized assessment in office buildings. *Energy and Buildings*, *153*, 136–149.

Armaroli, N., & Balzani, V. (2007). The future of energy supply: Challenges and opportunities. *Angewandte Chemie International Edition*, *46*(1-2), 52–66.

Asare-Bediako, B., Kling, W. L., & Ribeiro, P. F. (2013). Multi-agent system architecture for smart home energy management and optimization. In *IEEE PES ISGT Europe 2013* (pp. 1–5).

Asare-Bediako, B., Ribeiro, P. F., & Kling, W. L. (2012). Integrated energy optimization with smart home energy management systems. In *2012 3rd IEEE PES Innovative Smart Grid Technologies Europe (ISGT Europe)* (pp.1–8).

Avancini, D. B., Rodrigues, J. J., Martins, S. G., Rabêlo, R. A., Al-Muhtadi, J., & Solic, P. (2019). Energy meters evolution in smart grids: A review. *Journal of Cleaner Production, 217*, 702–715.

Balzani, V., & Armaroli, N. (2010). *Energy for a sustainable world: From the oil age to a sun-powered future*. John Wiley & Sons.

Bradac, Z., Kaczmarczyk, V., & Fiedler, P. (2015). Optimal scheduling of domestic appliances via MILP. *Energies, 8*(1), 217–232.

Chinchero, H. F., & Alonso, J. M. (2020). A review on energy management methodologies for LED lighting systems in smart buildings. In *2020 IEEE International Conference on Environment and Electrical Engineering and 2020 IEEE Industrial and Commercial Power Systems Europe (EEEIC/I&CPS Europe)* (pp.1–6).

Collotta, M., & Pau, G. (2015). A novel energy management approach for smart homes using bluetooth low energy. *IEEE Journal on Selected Areas in Communications, 33*(12), 2988–2996.

Fadda, G., Fadda, M., Ghiani, E., & Pilloni, V. (2019). *Communications and internet of things for microgrids, smart buildings, and homes. Distributed energy resources in microgrids* (pp. 243–273). Academic Press.

Ghasempour, A. (2019). Internet of things in smart grid: Architecture, applications, services, key technologies, and challenges. *Inventions, 4*(1), 22.

Hurtado, L. A., Nguyen, P. H., Kling, W. L., & Zeiler, W. (2013). Building energy management systems—Optimization of comfort and energy use. In *2013 48th International Universities' Power Engineering Conference (UPEC)* (pp. 1–6).

Ikpehai, A., Adebisi, B., & Rabie, K. M. (2016). Broadband PLC for clustered advanced metering infrastructure (AMI) architecture. *Energies, 9*(7), 569.

Jacoby, D. (2018). *From Bogotá to Beijing: Development and life after globalization*. Rowman & Littlefield.

Krishna, D. S., Rao, P. M., & Rao, T. S. (2021). *Emerging social information networks applications and architectures. Security in IoT social networks* (pp. 29–53). Academic Press.

Kuru, K., & Ansell, D. (2020). TCitySmartF: A comprehensive systematic framework for transforming cities into smart cities. *IEEE Access, 8*, 18615–18644.

Lorente S. (2005). *Domótica Integral: Análisis del Entorno*, Polytechnic University of Madrid, Master on Domotica, course material.

Mackenzie, A. (2005). The performativity of code: Software and cultures of circulation. *Theory, Culture & Society, 22*(1), 71–92.

Mahapatra, B., & Nayyar, A. (2019). Home energy management system (HEMS): Concept, architecture, infrastructure, challenges and energy management schemes. *Energy Systems*, 1–27.

Mahmood, A., Javaid, N., & Razzaq, S. (2015). A review of wireless communications for smart grid. *Renewable and Sustainable Energy Reviews, 41*, 248–260.

Mendes, T. D., Godina, R., Rodrigues, E. M., Matias, J. C., & Catalão, J. P. (2015). Smart home communication technologies and applications: Wireless protocol assessment for home area network resources. *Energies, 8*(7), 7279–7311.

Misleh, Y. (2014). *Homeostatic control of renewable micro-generation power systems: Towards a new approach to sustainable energy systems linked to energy efficiency*.

Mohammadi, F., Faghihi, F., Kazemi, A., & Salemi, A. H. (2021). A risk-based energy management system design for g rid-connected smart homes. *International Transactions on Electrical Energy Systems*, e12924.

Mohammadi, M., Noorollahi, Y., Mohammadi-ivatloo, B., Hosseinzadeh, M., Yousefi, H., & Khorasani, S. T. (2018). Optimal management of energy hubs and smart energy hubs—a review. *Renewable and Sustainable Energy Reviews*, *89*, 33—50.

Mohassel, R. R., Fung, A., Mohammadi, F., & Raahemifar, K. (2014). A survey on advanced metering infrastructure. *International Journal of Electrical Power & Energy Systems*, *63*, 473—484.

Moriarty, P., & Honnery, D. (2010). *Rise and fall of the carbon civilisation: Resolving global environmental and resource problems*. Springer Science & Business Media.

Olatomiwa, L., Mekhilef, S., Ismail, M. S., & Moghavvemi, M. (2016). Energy management strategies in hybrid renewable energy systems: A review. *Renewable and Sustainable Energy Reviews*, *62*, 821—835.

Pan, J., Jain, R., & Paul, S. (2014). A survey of energy efficiency in buildings and microgrids using networking technologies. *IEEE Communications Surveys & Tutorials*, *16*(3), 1709—1731.

Pascual, J., Sanchis, P., & Marroyo, L. (2014). Implementation and control of a residential electrothermal microgrid based on renewable energies, a hybrid storage system and demand side management. *Energies*, *7*(1), 210—237.

Pau, G., Collotta, M., Ruano, A., & Qin, J. (2017). *Smart home energy management*.

Qi, B. (2019). *Learning to control home batteries in the smart grid*.

Raja, S. B., & Baskar, N. (2012). Application of particle swarm optimization technique for achieving desired milled surface roughness in minimum machining time. *Expert Systems with Applications*, *39*(5), 5982—5989.

Rao, P. M., & Saraswathi, P. (2021). *Evolving cloud security technologies for social networks. Security in IoT social networks* (pp. 179—203). Academic Press.

Rathor, S. K., & Saxena, D. (2020). Energy management system for smart grid: An overview and key issues. *International Journal of Energy Research*, *44*(6), 4067—4109.

Saad al-sumaiti, A., Ahmed, M. H., & Salama, M. M. (2014). Smart home activities: A literature review. *Electric Power Components and Systems*, *42*(3—4), 294—305.

Schnieders, J., & Hermelink, A. (2006). CEPHEUS results: Measurements and occupants' satisfaction provide evidence for Passive Houses being an option for sustainable building. *Energy Policy*, *34*(2), 151—171.

Sepasgozar, S., Karimi, R., Farahzadi, L., Moezzi, F., Shirowzhan, S., Ebrahimzadeh, M., ... Aye, L. (2020). A systematic content review of artificial intelligence and the internet of things applications in smart home. *Applied Sciences*, *10*(9), 3074.

Shahzad, Y., Javed, H., Farman, H., Ahmad, J., Jan, B., & Zubair, M. (2020). Internet of energy: Opportunities, applications, architectures and challenges in smart industries. *Computers & Electrical Engineering*, *86*, 106739.

Shareef, H., Ahmed, M. S., Mohamed, A., & Al Hassan, E. (2018). Review on home energy management system considering demand responses, smart technologies, and intelligent controllers. *IEEE Access*, *6*, 24498—24509.

Shrestha, M., Johansen, C., & Noll, J. (2019). *Criteria for security classification of smart home energy management systems (long version)*. Research report. http://urn.nb.no/URN:NBN:no-35645

Sodiq, A., Baloch, A. A., Khan, S. A., Sezer, N., Mahmoud, S., Jama, M., ... Abdelaal, A. (2019). Towards modern sustainable cities: Review of sustainability principles and trends. *Journal of Cleaner Production*, *227*, 972—1001.

Stoyke, G. (2006). *The Carbon Buster's home energy handbook: Slowing climate change and saving money*. New Society Publishers.

Talari, S., Shafie-Khah, M., Siano, P., Loia, V., Tommasetti, A., & Catalão, J. P. (2017). A review of smart cities based on the internet of things concept. *Energies*, *10*(4), 421.

Tavakoli, M., Shokridehaki, F., Akorede, M. F., Marzband, M., Vechiu, I., & Pouresmaeil, E. (2018). CVaR-based energy management scheme for optimal resilience and operational cost in commercial building microgrids. *International Journal of Electrical Power & Energy Systems*, *100*, 1–9.

Tekbıyık Ersoy, N. (2012). *Efficient resource allocation in energy harvesting wireless networks*.

Usman, A., & Shami, S. H. (2013). Evolution of communication technologies for smart grid applications. *Renewable and Sustainable Energy Reviews*, *19*, 191–199.

Vandermolen, S. J., & Cella, E. P. (1999). A safety, health and environmental management system (SHEMS) approach to capital projects. In *Offshore Technology Conference. Offshore Technology Conference*.

Vasileiadou, E., Huijben, J. C. C. M., & Raven, R. P. J. M. (2016). Three is a crowd? Exploring the potential of crowdfunding for renewable energy in the Netherlands. *Journal of Cleaner Production*, *128*, 142–155.

Xue, B. (2017). *Distributed communication middleware for an selector model* (Doctoral dissertation, Rice University).

Yamashita, D. Y., Vechiu, I., & Gaubert, J. P. (2020). A review of hierarchical control for building microgrids. *Renewable and Sustainable Energy Reviews*, *118*, 109523.

Yang, Q., Li, J., Cao, W., Li, S., Lin, J., Huo, D., ... He, H. (2020). An improved vehicle to the grid method with battery longevity management in a microgrid application. *Energy*, *198*, 117374.

Yang, T., Zhao, L., Li, W., & Zomaya, A. Y. (2020). Reinforcement learning in sustainable energy and electric systems: A survey. *Annual Reviews in Control*, *49*, 145–163.

Yang, X., Zhang, Y., Zhao, B., Huang, F., Chen, Y., & Ren, S. (2017). Optimal energy flow control strategy for a residential energy local network combined with demand-side management and real-time pricing. *Energy and Buildings*, *150*, 177–188.

Zhou, B., Li, W., Chan, K. W., Cao, Y., Kuang, Y., Liu, X., ... Wang, X. (2016). Smart home energy management systems: Concept, configurations, and scheduling strategies. *Renewable and Sustainable Energy Reviews*, *61*, 30–40.

Chapter 4

Role of power line communications in the Smart Grid: applications, challenges, and research initiatives

P. Muralidhara Rao and B.D. Deebak
School of Computer Science and Engineering, Vellore Institute of Technology, Vellore, India

4.1 Introduction

Energy resources are considered prime factors for the socio-economic growth of the nation. It should join the hands in a more sustainable way given societal, economic, and environmental aspects. Like other developed countries, India is also energetically working on the development of modernizing power grids. India is known for its second-largest population and seventh-biggest geographical area. In India, providing 100 percent electricity across the nation is becoming a challenging issue due to supply uncertainity in fuel and coal mining. However, the Indian electricity utility system (EUS) overcame the challenge in 2018 (Schmalz et al., 2018). The EUS providers currently focus on achieving new goals like commercial cost, electric power quality improvement, increased power supply reliability, fault reduction, efficient power transmission, and real-time electricity tariffs. Enabling all these features could enhance the efficient use of primary energy resources like fuel, natural gas, and coal, achieving customer satisfaction, improved industrial growth, and economic development.

Notwithstanding the extensive growth of electric power utilization in the last decade, it necessitates the growing concern regarding power quality, effective power supply, availability, and cost-effectiveness. The SG can be a promising solution to provide enhanced features and advantages to increase the operational and technical efficiency of the existing electric grids (Oliveira, 2015; Zaglago et al., 2013). The SG was developed to improve the power transmission and operational efficiencies. It integrates advanced communication technologies to guarantee higher reliability, flexibility, efficiency,

Sustainable Networks in Smart Grid. DOI: https://doi.org/10.1016/B978-0-323-85626-3.00001-6

greater robustness. Besides, it allows optimum use of renewable energy resources resulting in a greenhouse gas emission reduction (Colak et al., 2016). So as to accomplish the goals of SG, several aspects of the power grid, including generation, transmission, distribution, and other related electric equipment, must be upgraded. Then the advanced control and monitoring systems can adopt SG's features to ensure hassle-free services.

Nowadays, ICT provides remarkable services in the development of industrial and socio-economic applications for the wellbeing of 21st-century humans. The rapid growth of these ICT services brings numerous advantages over the application development and proper utilization of Internet resources (Chant & McIlwaine, 2009). Therefore, ICT can be a key concept for developing SG applications specific to SM and advanced metering infrastructures (AMI) infrastructures. Besides, power line communications (PLC) provides numerous advantages, particularly to these SGs and their related applications, that it extensively uses conventional electric grids as the communication method. Achieving the providence for the enabling technologies for SG's development has been an active research topic in the last few years. It is still open to researchers to find advanced solutions to enhance the robustness of the grid (Colak et al., 2015). The PLC technology can effectively work only when they have efficient bandwidth for effective communication. The PLC can be primarily classified according to the bandwidth, range, and data rates. It can be classified into three types: Ultra narrowband PLC, Narrowband PLC, Broadband PLC, and Quasi band PLC (Yigit et al., 2014).

Conventional power distribution systems increase voltage levels to transport electricity from power plants, then gradually reduce voltage levels to supply electricity to end-users. The use of real-time pricing and the growing relevance of renewable electricity generation necessitate intelligent technologies in the electricity market. The smart grid is a way to ensure user safety by adding intelligent meters and monitoring devices to the electrical grid allows for continuous monitoring, upgrading, and distribution to the power grid to assure electronic connection between suppliers and customers. The distribution of good intensity to consumers is a key success for the smart grid. The SG framework will enable the use of digital information technology to defend against attacks, allow for customer participation, optimize power supply and generation, deliver high-quality power, support different generation options, and self-heal. The generation domain shares information with the electricity market, the control center, and the regional system operator are all involved in this process.

As a result, in the event of a lack of generation or a generator breakdown, the power market and the regional system operator make immediate plans. Dynamic pricing is predicted to be affected by the integration of new generation resources. There is a surge in technical development of residential structures to become smart as smart home components and IT integration become more widely used. To get the most out of real-time pricing and maximize renewable strength resources (RER) use, smart components must be properly planned. However, to investigate the impact on operational costs and practical aspects, researchers did not include the cost of battery deterioration in the demand-scheduling program. In the authors' optimal scheduling of storage devices, the battery deterioration cost was employed as the objective function. Therefore, this chapter provides comprehensive detail on the role of PLCs in the growth of SG technological adoptions for real-time solutions.

4.2 Theoretical background

4.2.1 History and evolution

In the early ages of communication technology, the first PLC applications were used for power utilities by involving voice and data communication through high-voltage (HV) power lines capable of handling more than 100 kV and serving large geographical areas. PLC is an old idea that serves electric utilities for remote metering and load control across the world. Besides, HV lines (power carrier systems) have been used for voice communication for decades from the early 1920. In the early 1900s, the telephone coverage was not enough to transmit, whereby engineers used PLCs as an alternative solution for enabling power plants and transformer stations (Ferreira et al., 2011). Later on, digital communication technology was introduced. However, it servers extremely limited range and limited bandwidth up to Mbps.

The first single carrier NB was used to operate low-frequency audio services that achieved data rates from extraordinarily low bps to a kbps. As emerging technology brings new advancements, the application space also increases to serve numerous advanced services. BB-PLC systems were introduced to operate on high-frequency levels ranging from 2 to 30 MHz to achieve a data rate up to 200 Mbps (Ndjiongue & Ferreira, 2019). Of late, excessive expansion of industry requirements and its interest has grown across the high data rates. Lately, the NB-PLC was introduced based on a multicarrier system operating on extremely high frequencies, that is, 3–500 kHz. Moreover, PLC-based technologies can offer BB internet access for domestic use, as well as home, workplace, and car monitoring and control, as well as remote monitoring and control for autonomous metering services (Huh, 2017a, 2017b). The main advantage of using PLC technologies in the power grid delivers an extensive infrastructure that is ubiquitous than existing solutions.

Despite the promising technology that enables various solutions in the applications of electricity utilities in both now and in the future, PLC technology has not yet been accomplished the majority of market requirements with its ability. However, the most compelling rationale for employing PLC is to modernize management of the power grid for handling energy transmission and distribution. It is mostly believed that the mass expansion of power utility demand has exceeded, and many governments agreed to minimize the greenhouse gas emissions to prevent climate change (Huh & Park, 2020). Finally, it is necessary to modernize the power grid infrastructure across the world with minimum investments. Another important aspect of PLC is load control, which can control on/off procedures for equipment that consumes a lot of energy, such as air conditioners, water heaters, and other household items.

Ripple carrier signaling was a successful initiative in well-developed countries like Europe to the demand side, control peak occurrences by permitting control signals to operate on heavy-duty appliances (Ourahou et al., 2020). PLC has a long history serving electrical businesses with a diverse set of applications and implementations with the help of telecommunication technology. The merging of smart metering and PLC technologies have brought remarkable services in the field of electrical utilities. The plan of bringing PLC into SG technology can be recognized as tremendous technology which can support broad range of applications and services to the long-range and home appliances ranging from transmission to distribution (Borlase, 2017). Though, the application of PLC to the SG is not directly connected, the learning curve for PLCs may be difficult for the electricity company that require the necessity of PLC to deploy a service-oriented system. The particular details are never discovered in the traditional electric grid. Therefore, the deployment of PLC solutions can be highly essential for the adoption of real SGs. Fig. 4.1 depicts the worldwide usage of energy growth can be anticipated between 2007 and 2050 (Ourahou et al., 2020).

The SG European Technology Platform defines SG as "an electricity network that can automatically incorporate all users' behaviors" linked to it—generators, consumers, and those who do both—to assure a better long-term, cost-effective, and secure electricity supply (Mutoh et al., 2006). SG in a digital technology, according to the US Department of Energy, is used to improve the electrical system's dependability, from large generators to power users, security and efficiency (both economic and energy) are paramount. A growing variety of distributed-generation and storage resources are now available. The goal is to anticipate the upcoming, including supply network futures, to generate tomorrow's solutions, engage with and aid communities at the local and territorial levels, and to innovate and adapt new technologies to put it another way, it's about aiding the territories' economic and social growth, as well as easing the energy transition, by doing things like:

- Allowing the network integration of renewable energy;
- Accompanying electric vehicle development and;
- Encouraging the influenced contemporary renewable energy techniques and evolution of the electric car.

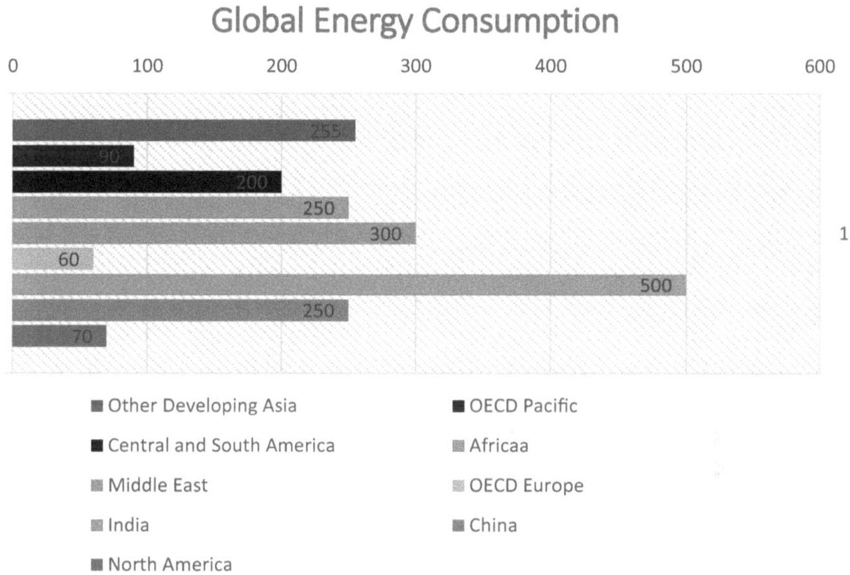

FIGURE 4.1 Global energy consumption.

While enhancing the power supply's quality and minimizing routing costs, these three issues must be addressed via the distribution network (renewable energy, electric vehicles (EVs), and energy demand control). The energy distributor should serve as a facilitator as well as a federator of energy. Indeed, the SG adventure, the power of digitization, necessitates the grouping and federation of various expertise and partners.

4.2.2 Characteristics and advantages

Most private residences lack dedicated low- and high-speed network cable, and the labor expenses to install such wiring are frequently prohibitive. PLCs communicate using the existing electrical network. As a result, installation costs are lower than for other communication systems, and communication service is available wherever outlets exist. Remote measurement and control task low, medium and high voltage stockpiles were employed for effective operations of electrical utilities (Lambrechts & Sinha, 2019). PLC is often utilized for numerous communication applications in internal electrical installations within buildings and houses, referred to as in home PLC. With high data transfer rates of up to a few hundreds of megabits per second and house power line architecture depicts an appealing network for sharing data among intelligent gadgets, including electrical outlets in each room.

The electricity grid must be modernized to handle energy shifts. Designers and specialists throughout the world are working to improve electrical networks by deploying SG technology rather than replacing and massively reinforcing the grid. The grid will become more communicative as new ICTs have been integrated into it, allowing a wide range of participants to engage while providing more effective, economically feasible, and safe electricity distribution in new forms of power systems, such as EVs, from one in which legitimate equilibrium is achieved by adapting output to consumption, to one in which demand adjusts more, turning the consumer into a true actor. Corporate environments that are more demanding, combined with the advancement of technology, has resulted in the Smart Grid. Systems that were previously unaffordable items are now widely available. As a result, the electrical system will be more flexible in dealing with constraints like renewable energy's intermittent nature and development.

Of late, SG delivers a promising enterprise solution which could ensure the far-reaching advantages for both utilities and their customers. Utilities that adopt SG technologies can reap tremendous advantages in operational costs and reduces capital, improves power quality, customer satisfaction, and positive environmental impact. The solutions delivered from SG not only based on enabling solutions but also addresses business and operational concerns. It assures to deliver promising solutions in order to deliver measurable, meaningful, sustainable advantages to the consumers, utilities, environment and economy. Fig. 4.2 depicts the advantages of SG in a general sense.

Several components can join their utilities for delivering numerous services considering SG technologies. Besides, utilities and consumers can get the major benefits out of delivering major technological services with integration of SG. It facilitates increased utility earnings, reduced rate of capital, and increased quality of service.

New PLC technology offers excellent due to the advent of new PLC technology, characteristics and advantages in types of sensory processes,

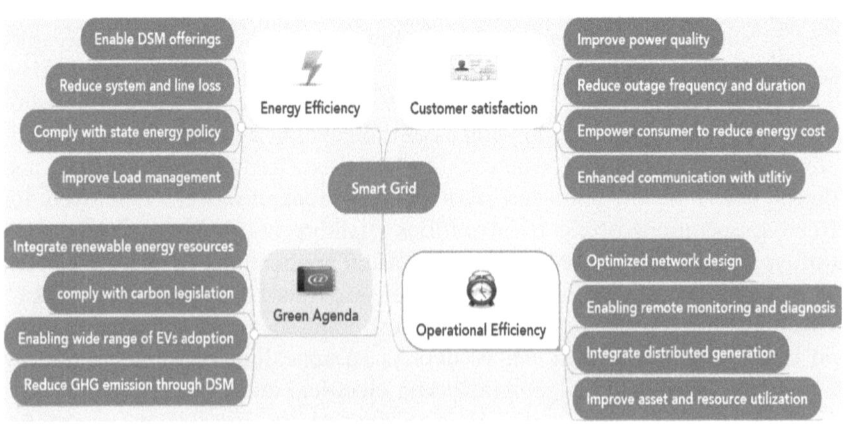

FIGURE 4.2 Advantages of the Smart Grid.

analogue source and destination, human-machine switching, and network connectivity are all examples of input and output. It has truly recognized the shift from analogue to digital control, and its application area and focus are growing. It is also used to give integrated control in solar hot water engineering and construction, making the control mode more user-friendly. In other cases, the roof slab functions as both the building's sloping roof and the collector plate for the solar system. In order to encourage solar and building component collaboration.

4.3 Power line communication: overview

4.3.1 Role of power line communication

SG has gained numerous advantages and serves remarkable services in the development of sustainable energy management. It gains a lot of attention, presenting its inherent capacity to ensure better performance utilizing intelligent grids for future results. Besides, it requires a bidirectional communication platform between consumers and producers to enhance the electrical transmission system's productivity, safety, reliability, and energy efficiencies. In this concern, PLCs have been an alternative solution, which assures the end-to-end communications starting with the most basic of domestic applications to the largescale industrial applications (Sharma & Saini, 2017). Fig. 4.3 depicts the overview of the PLC-based SG power system. It enables monitoring and control-based services to distribute generated management applications with reliable network access. PLC's offer various services with

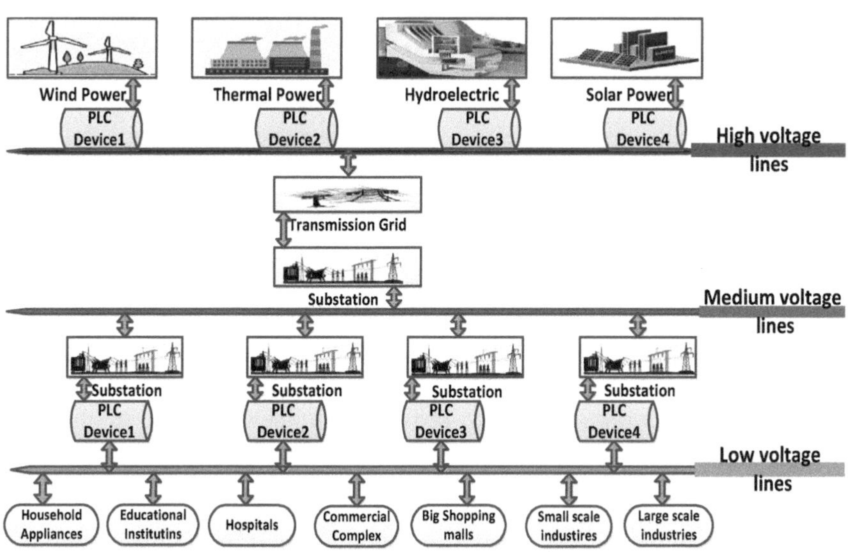

FIGURE 4.3 PLC-based SG infrastructure architecture.

the help of different communication technologies, namely, GPRS (wireless access), UMTS (mobile communication technology), RF (Radio Frequency technology), and WiMAX (worldwide interoperable service). The PLC brings numerous advantages for the betterment of SG (Roslan et al., 2019); they are as follows:

1. Standard communication establishment during the installation can assure minimized wait time for the installer.
2. No need tuning or adjusting the antenna is required for antenna direction.
3. Reliable communication can be possible in hostile environments like deep undergoing installations, meta-shield cases, etc.
4. Antenna protection can be done using this PLC; for instance, antenna tampering protection can ensure reliability, which cannot be failed.
5. The PLC interface can be equipped with a 230 V power system ensures security against tampering.

The rapid advancements and excessive usage in electricidal utilities have raised investments in SG infrastructure worldwide. As a result, the PLC's global market can be expected to be more than 7.5 billion (USD) by the end of 2022. The present subsurface cabling systems and SG overhead wires necessitate the introduction of PLC technology at a low cost (Ponce-Jara et al., 2017). The rapid advancements, new standards, advanced regulations, and certifications in PLC technologies have raised the growth estimated for the worldwide market in several end-user applications. These applications include AMI, remote telemetry systems, vehicle automation, equipment for high-speed internet access, smart home equipment, electric vehicle charging, largescale industrial systems, end-to-end communications, and microinverter services for alternative energy and temperature monitoring and lighting.

However, the excessive use of increased certifications, standardization of PLC technologies, enhanced connectivity in openness in communication technologies in the SG has opened doors to adversaries so that it is highly vulnerable to various attacks and security threats on PLC-based hardware equipment and software protocols (Faheem et al., 2018; Knapp & Samani, 2013).

4.3.2 Infrastructure: traditional to Smart Grid

The primary purpose of SG is to furnish the electricity network with the tools it requires to become intelligent. For supply security concerns, the transportation network is already heavily instrumented. Due to the vast amount of equipment (stations, wires, etc.) and customers linked to these networks; however, they are under-equipped with communication technology. As a result, distribution networks are important to smart grids (Al-Badi et al., 2020). With the advanced technologies smart electric utilities have become emerging because they incorporate information and communication technology, smart grids are communicative. The ability to examine the behaviors of various characters in the electric grid, notably consumers, is made possible by this connection between the various locations of the grid. The goal is to follow a standard supply and demand balance while improving responsiveness and dependability and optimizing network functioning. The electrical system evolves from a linearly operating chain to a system in which all actors are involved [ref]. Figs. 4.4 and 4.5 depicts the various actors who contribute to the complexity of conventional and smart grid systems. The characteristics of conventional and Smart Grid is depicted in Table 4.1. In actuality, there are three levels to smart grid architecture as described below:

- The first level uses normal infrastructure to deliver electricity (transformers, and lines, etc.).

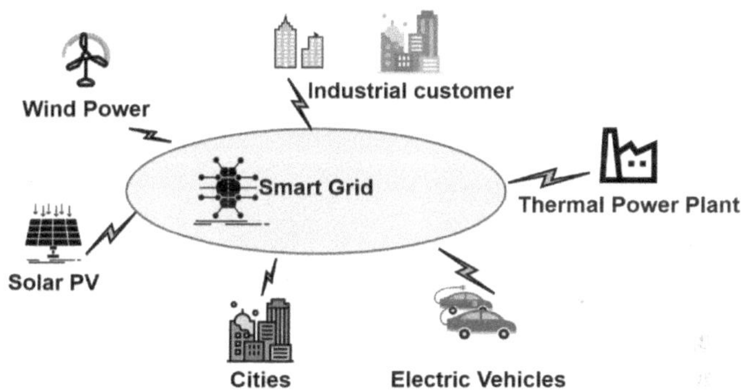

FIGURE 4.4 Smart Grid.

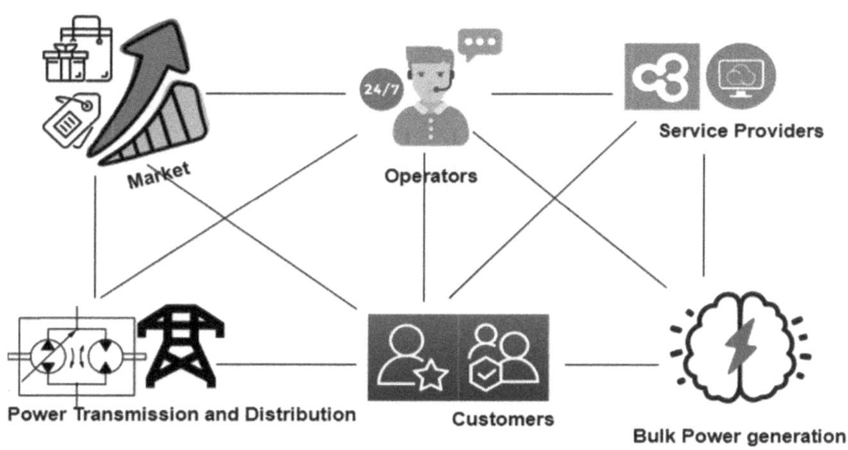

FIGURE 4.5 Conventional Grid.

TABLE 4.1 Characteristics of conventional and Smart grid.

Properties	Conventional Grid	Smart Grid
Signal	Analog	Digital
Path	Unidirectional	Bidirectional
Production	Centralized	Decentralized
Communication	Some parts of the grid	All over the grid
Management	Supply-side electrical system equilibrium	Demand-side electrical system equilibrium
Utilization	Versus production	Versus consumption

- The level 2 is a communication infrastructure that employs a variety of media and technology to communicate (fiber optics, GPRS, PLC, etc.) to gather information from sensors installed on electrical networks.
- The level 3 includes remote troubleshooting systems and demand response algorithms that use real-time data are examples of applications and services.

4.3.2.1 *Market*

With the emergence of SG, a brand-new market is emerging. The energy flow is being affected by providing new services and selling brand-new items. The emergence of a new market based on energy trading and the provision of auxiliary services follows. Hourly power purchase and sale contracts can be entered by both producers and consumers.

4.3.2.2 *Operators*

The two types of operators are essential for SG communication that is, distribution and transmission. They are both are too accountable for the transmission and distribution grid's reliability, efficiency, and smooth functioning. They must evaluate and operate a vast volume of data from many sources in real time, as well as provide a variety of services include Smart metering, energy storage, plug-in electric vehicles, and plug-in hybrid electric cars are all examples of demand/response management.

4.3.2.3 *Consumer*

In a SG system, customers could play an active role in trade of energy by either utilizing or providing energy. A new leadership profile has been adopted to maintain demand/balancing the response, and addressing supply issues, and encourage customers to utilize low-energy appliances for weekends and evenings are off-peak hours.

4.3.2.4 Bulk power production

It encompasses both variable renewable energy sources and centrally managed traditional power facilities. To maintain grid stability and absorb the influx of intermittent renewable energy sources, energy storage is critical.

4.3.2.5 Service providers

It allows businesses that provide commercial services like billing and charging, as well as technical services like constancy, security, energy control, and residential power generation. It serves as a link between the customer, market and the service providers. All grid-related services are available for the industrial business can be performed by service providers.

4.3.2.6 Transmission and distribution

On just one side, dealing with bidirectional energy streams on a grid that was designed largely for linear demands are a challenge for a distribution network. Furthermore, it must connect all micro-grids and small-distributed energy sources into a smart metering-enabled technologically sophisticated grid. The transmission grid, on the other hand, must maintain a dependable a link between large-scale power generation and the distribution system that considers both economic and technological considerations.

4.3.3 Communication standards

PLC is a communication technology that facilitates transmitting data over existing power cables. With the help of power cables fixing to an electric device, it can power up and retrieve the data from it in a half-duplex mode. According to the market strategy, PLC technologies have broadly classified into four categories:

1. Ultra-Narrowband PLC (UNB-PLC)
2. Narrowband PLC (NB-PLC)
3. Broadband PLC (BB-PLC)
4. Quasi band PLC (QB-PLC)

Narrowband PLC segments work at a lower frequency, that is, 3–500 kHz, lower data rates 100 kbps, and kilometers longer-range, and it can be expanded using repeaters. Narrowband PLCs have attracted widespread attention due to their massive usage in SG infrastructure. Narrowband PLC can also be used in smart energy generation, especially in microinverter for solar panels, an alternative energy source (Fenton, 2006).

In contrast, broadband PLC operates at a higher frequency, that is, 1.8–250 MHz, high data rates, that is, more than 100 Mbps, and is used in short-range applications, that is, below kilometer. Therefore, it mainly serves as a promising technology for internet distribution and home networking. With the benefits of higher data rates and lower wiring, broadband PLCs are becoming an effective solution for multimedia distribution technology for home networks. These features bring optimistic ways of attracting the market acquisition of various lead companies (Trl et al., 2014).

Alongside, PLC can be classified in another way is as follows:

1. PLC over AC line
2. PLC over DC line

Nowadays, most of the competitor companies pitched towards the development of AC line DC line PLC applications. Over DC, these PLC technologies can be used in electronic controls in planes, trains, and other vehicles, distributed energy generation in the transportation segment (Dell & Rand, 2001). PLCs in these segments can minimize wiring complexity, weight, and communication cost over the terminals and inside vehicles (Table 4.2).

4.3.3.1 Power-line telecommunications

It is a European Telecommunications Standards Institute (ETSI), establishes the essential voice and data transfer standards over electricity lines. The Home-Plug Power-Line Alliance is a non-profit organization based in the United States with 65 companies as members (ETSI). The major goal is to make home power-line networks that are cost-effective, interoperable, and

TABLE 4.2 Communication technologies for Smart Grid.

Technology	Merits	Demerits
• Optical fiber	• High capacity • Stable characteristics	• High cost
• Wireless	• Cost-effective • Mature technology • Rapid installation	• Limited coverage • High capacity • Security • Reliability
• Satellite	• Global coverage • Rapid installation	• High cost • Long delay • Dynamic channel
• PLC	• Extensive coverage • Available Infrastructure • Reliable • Cost-effective	• High noise • Signal Attenuation • Cable loss and multipath • EMI

based on industry standards and solutions more readily available, adoptable, and implemented. The standards that result are anticipated to deliver the finest results possible Some criteria have been classified by the Home Plug Power-Line Alliance, including (Galli & Logvinov, 2008; Galli et al., 2009).

First, House Plug 1.0 is a standard for using electricity lines to connect gadgets in the home.

Second, Home Plug AV—a specification for transferring high-definition television (HDTV) and VoIP throughout the home,

Third, Home Plug BPL—a working group tasked with creating a standard for to-the-home connections. and

Fourth, The Home Plug Command and Control (HPCC) standard allows for enhanced whole-house lighting, appliance, climate automation, security, and other device control.

4.4 Power line communication's: advanced technologies and applications

4.4.1 Smart metering

Smart Grid has gained research interest over the decade, widely used in energy distribution utilities with the help of Smart Metering [Ref] services. SM is also belongs to Automatic Meter Reading and the infrastructure related to Automatic Metering Infrastructure (AMI). SM use is more than traditional remote metering abilities to operate on smart meters, which becomes a new service for energy-efficient mechanisms (Kemal, 2019). Besides, several new services have been used in the wider SG electricity distribution for advanced grid control requirements and utilities. SM's require both basic functionalities needed for remote metering, and real-time metering requires real-time access for connection and disconnection. Consequently, the SM technology creates the necessity to work close to the SG extended functionality to develop the smart infrastructure capable of increased outcome through distributed networks (Avancini et al., 2019).

The SM provides numerous services that are in line with the development of SG, namely, Grid Modeling, Design Systems, Simulation, Demand Management System (DMS), Asset Management System (AMS), Intelligent Network Management System and Agents (INMSA), Distribution Autonomous System (DSA), Energy Management System (EMS), and Geographic Information System (GIS). These advanced services are mainly used to provide smart and remote services. For instance, DMA is used for remote substation control. EMS is used to collect to assess, analyze, manage, and plan the use of SG using data from numerous sensors, smart meters, and other functional devices. DMS is used to reduce consumption and optimize the global performance of the SG. AMS includes the utilization of an updated inventory system. GIS enables the geographic data to provide optimized planning activities, detection, and fault recovery systems. Similarly, INMSA provides intermediate data collection from various grids, facilitates interaction between control centers to take dynamic decisions on the local control functions.

4.4.2 Automatic/advanced metering infrastructure

The AMI can design a capable system in handling both collections and manage data from SM's to enable bidirectional communication. It performs three operations: initially, it allows services to measure real-time, detailed information. Second, it collects all the data from various sources. Lastly, it transmits data to the specific device in the network (Mohassel et al., 2014). Utilizing AMI services ensures the consumers with advanced services and products for minimized quality of services for each region based on their country regulations. Consequently, productivity can be increased so that empowerment can be expanded for the involvement of stakeholders in the electricity domain.

4.4.3 Home automation

Of late, AMI has become a critical foundation because of its two-way communication and always-on connectivity capabilities, and it has been ideal for the smart home uses of today. The smart home energy management system is a typical application that does the following:

1. Obtains energy-related data from the smart meter (such as energy consumption and real-time price information),
2. Stores the energy-related data and provides power usage statistics, and
3. If necessary, it can control smart appliances via home communication networks.

However, security concerns in AMI-enabled apps remain an open topic and a key problem. SG incorporates breakthroughs from transdisciplinary technologies to facilitate significant power grid monitoring, management, operation, and control. At the feeder level, Distribution Automation solutions are used, and at the LV distributor level, AMI-based solutions are used, smart grid capabilities can dramatically improve the distribution system's reliability, performance, and efficiency.

Distribution automation systems include Fault Detection, Isolation, and Restoration, Feeder Reconfiguration, and Integrated Volt-Var Control (Schwarzenegger, 2007), which improve efficiency and assure electricity distribution reliability and quality. Designating applicable meter supply disconnect and reconnect, load control, meter loss of supply and outage detection, power quality and events recording, supply capacity control, and providing interface to HAN, tamper detection, communications, and data security, and firmware for AMI single and three phase smart meters.

4.4.4 Home networking and internet access

Demand Response (DR) programs allowed by the AMI in the interests of both utilities and residents, were incorporated into the electrical grid, integrating power grid technology with renewable power producing technologies. Residents are encouraged to limit during high load hours, they use a lot of energy in exchange for incentives, which helps to achieve load balance and increase system resilience. To automate this method, the AMI controller, as well as all appliances, in-house sensors, and the AMI controller, must all be connected. Both wired and wireless network technologies ideal for deployment in houses must be compared and evaluated (the central SHEMS Smart appliances, sensors in the home, energy generators, electric vehicles, and other devices are all linked.) for SHEMS to work in a smart home (Siano, 2014).

Due to a shortage of realistic test-beds with a sufficient scale for independent tests, as well as software simulation environment assistance, it is difficult to conduct independent experiments capable of facilitating few papers and articles exploring such concerns, particularly in terms of energy management, as a result of such evaluations. Furthermore, in the goal of energy management on a broad scale, from the standpoint of the overall grid, more emphasis was placed on energy management and its infrastructures supporting communication Industry offers a networking system that combines PLC and wireless communication for residential usage, rather than locals, to close this gap and promote additional smart dwellings, research into energy management and networking technology is underway (Throndsen & Ryghaug, 2020). With the help of AMI, DR program utilities may achieve load balancing in the power grid by alerting consumers about the dynamic characteristics of supplying power to them and encouraging them to use as little electricity as possible during high demand periods.

Smart meters in smart homes offer energy management recommendations to smart appliances after receiving time-varying pricing messages via DR procedures. As to achieve end-to-end pricing across a wireless or wired network to transfer and power consumption adjustments with the goal of lowering energy consumption and increasing efficiency. Authors in (Jin & Kunz, 2011a) provides an operational simulation model in NS-2, merging PLC technology and wireless technologies to examine network traffic and energy consumption in smart homes is DR message transmission based on this concept. A Radio Broadcast Data System (RBDS) network is included in the simulation model, as well as an indoor networking scheme that combines combination of ZigBee/IEEE 802.15.4 with HomePlug C&C creates a situation in which there are several interfaces (up to three) and channels. A smart home controller has three interfaces: one to the RBDS network, one to the internet, and one to the local network. The electricity utilities' (or third-party service providers') RBDS message sender has just one interface to the

RBDS network. One is linked to the ZigBee/IEEE 802.15.4 network, and the other to the HomePlug C&C network (Jin & Kunz, 2011b).

The ZigBee/IEEE 802.15.4 and HomePlug C&C networks both include interfaces to the intermediary nodes, which are spread throughout the house. They deliver RBDS messages around the network based on the routes specified to destinations, acting as both message recipients and routers. These nodes can also be configured to solely work through the HomePlug C&C network's interface, simulating smart home appliance behavior connected to the power line. The remaining nodes, which each have a single ZigBee/IEEE 802.15.4 network interface, function as in-home sensors or general-purpose smart home appliances can be put anywhere in the house and receive RBDS signals on an irregular basis.

4.4.5 Vehicle to Grid

Of late, the electric vehicle market has grown exponentially. According to the International Energy Agency's Global EV Outlook 2019 (Kampker et al., 2021), the number of self-contained electric vehicle charging stations that are typically linked to low-voltage distribution networks, increased by 48% over the previous year to 4.66 million in 2018. Light and medium-duty vehicles, as well as electric passenger automobiles, are becoming more popular across the world, and this trend is expected to continue for EV chargers consume a lot of power even at low voltage. This makes maintaining the distribution system's reliability difficult, especially in areas where EV penetration is high (Muratori et al., 2021). The high EV charging demand has transformer overloading, voltage sag, greater losses, and shorter equipment lifetimes ensued. Although updating the distribution network can address these issues, the huge expenditures needed may make it impractical. Another alternative is to take advantage of EV charging loads' inherent flexibility to organize them to use available network capacity without generating these issues. In related study, two types of EV charging control algorithms have been presented: centralized and decentralized.

A central controller determines the first category of the maximum charging power for each connected After evaluating a number of factors, EV was chosen. After that, the transmission of charge powers to the appropriate charging stations. Because all data about electric vehicles and other metrics must be transmitted, these systems have a large communication cost and do not scale with the network's size or number of charging stations.

Instead of using network analysis to simulate if there is congestion in a specific area of the network, a feedback control algorithm the design has inspired me of digital traffic delays control algorithms could be used to infer congestion (i.e., overloading and voltage problems) from local or remote sensor measurements. For example, charging stations can alter their power in response to feedback, ensuring that overall EV charging demand matches network capacity in real time (Al Zishan et al., 2021). These algorithms are based on the Additive-Increase Multiplicative-Decrease (AIMD) rule-based control system, which has been successfully implemented in TCP. TCP congestion management has been adopted. Regardless of network changes, AIMD's control rules remain constant (Lorincz et al., 2021). Fixed rules are not a huge concern for short-term control, but they might cause inefficiencies in decision-making. Fixed rules are not a problem for short-term control, but they can cause inefficiencies when making decisions on a longer period, as seconds in the distribution grid.

4.4.5.1 Algorithms for electric vehicle charging congestion control

Researchers were inspired by the application of TCP congestion control techniques to power system congestion management has been a success. The authors of (Al Zishan et al., 2021) suggest a decentralized charging technique to maximize EV charging capabilities based on the AIMD algorithm while retaining fairness Raising When the network becomes congested, the charge rate of an EV is multiplicatively lowered after being enhanced by an additive factor. They do not, however, take into account practical power system constraints such transformer ratings and voltage limits.

The authors of (Zishan, 2020) extend the AIMD technique of (Al Zishan et al., 2021) to include these requirements while assuming a communication network with a hierarchical structure. The authors also offer a heuristic for promoting EV charging during off-peak hours that is price-adjusted available power. Liu and McLoone (2015) just published a fully decentralized AIMD-like EV charging scheme. This approach is based primarily on local voltage measurements and capable of being fine-tuned to get close to it. Table 4.3 depicts several crucial applications for the SG with their communication requirements and other specifications. The communication requirements follow based on data rates, size and latency (Alam et al., 2017). optimal performance. Fine-tuning the parameters, on the other hand, is not easy. Because the approach may not be able to completely utilize the network's available capacity if it relies on local data. Furthermore, while assigning power to EV, the approach does not consider fairness.

4.5 Power line communications for the Smart Grid

PLC technology is a significant aspect of computer technology in general. It has the ability to monitor and control the entire industrial production process to guarantee that it runs smoothly. The following three components of PLC technological qualities are the most important. PLC technology, for starters, offers a powerful anti-interference capability as well as a self-diagnosis feature. When inaccurate data is detected, it can be identified and corrected in

TABLE 4.3 Smart Grid applications and specifications based on communication requirements.

Application	Data range	Latency	Task performed	Data size (Bytes)	Reliability (%)
Remote control	~bps	>100 & <500 ms	Connections and disconnections	100–500B	>99
SCADA	~kbps	300–3000 ms	Monitoring and control	100–500B	>99
Lighting and traffic control	~bps	100–300 s	Signaling	10–50B	>99
Cost control	~bps	10 s	Pricing/Payment related services	10–100B	>99
Substation automation	<56 kbps	15–200 ms	Monitoring and control	150–250B	>99.99
Domestic applications	~kbps/device	<15 s	Customer information request and response	100–500B	>99
Distribution automation	<56 kbps	15–200 ms	Fault detection, clearance, restoration and isolation Monitoring and data maintenance	25B, 100–1000B	>99.99
Home energy management	<56 kbps	200 ms–2 s	Signaling	100–500B	>99–99.9

Distribution management	<100 kbps	100 ms–2 s	Monitoring and data maintenance	100–1000B	>99–99.99
Distributed energy management	<56 kbps	200 ms–2 s	Operation commands	25B	>99–99.99
Outage management	56 kbps	2 s	Outage and restoration management	25B	>99
Demand side management	14–100 kbps/node	2–5 min	Service switch operation	25B	>99
Electric Vehicle to grid	<56 kbps	2–5 min	Pricing signals and commands	255B	>99–99.99
Surveillance	500–1500 kbps	~s	Monitoring and response	100–500B	>99.99
Automated distribution automation	~Mbps	25–100 ms	Monitoring, control, and protection	>55B, 4–160B, 4–160B	>99.99

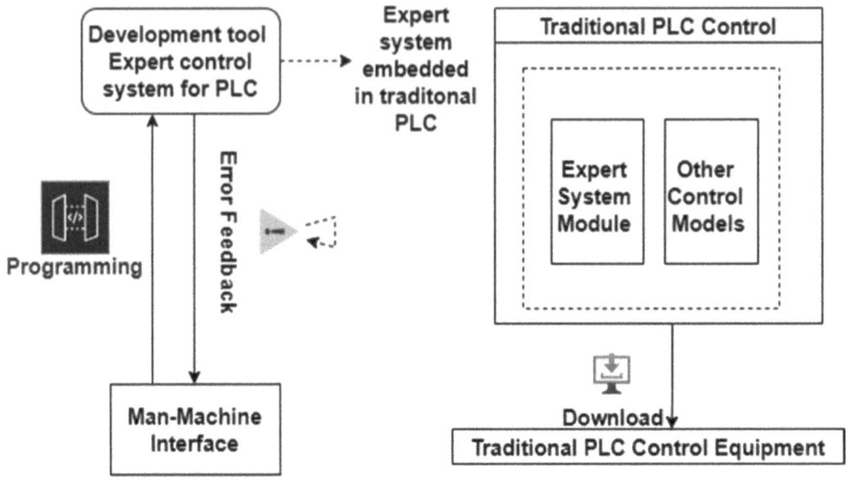

FIGURE 4.6 PLCs operational logic (Zhao & Zheng, 2021).

real time utilizing methods like filtering and isolation. PLC technology can demonstrate excellent stability and reliability in real-world applications, allowing it to acclimate to a more difficult working environment is depitcted in Fig. 4.6.

Second, the functions of the PLC system is fairly complete. In the meantime, it is highly adaptable and can be used in conjunction with industrial manufacturing. It may function in a complicated industrial setting, expand system functions, and vary the degree of control, among other things. Third, ladder diagrams are used to program PLC technology. This strategy is simple to understand. In a short period of time, the relevant personnel can learn the operation procedure. The wiring workload is considerably decreased when the channel is replaced by software that is related. Furthermore, the staff is not required to evaluate the issue of node relocation. If you need to make changes to a program, you can do it on the hardware or software directly on the other hand, such software consumes little energy and is relatively small in weight. Even if the necessary staff does not have extensive computer skills, they can perform a variety of everyday tasks. Fig. 4.1 depicts the unique PLC technology operating logic diagram.

4.6 Power line communications: applications and real-time solutions

4.6.1 Coal mining

The extraction of coal resources necessitates the building of a coal resource mine the usage of numerous mining equipment's. When compared to true mining equipment, automation, and management, it is more advanced programmable control technology. When it is utilized, it may be modified by a variety of conditions, posing new safety risks. As a result, programmable control technology can be applied to the digitization of mining gear in order to enhance mining equipment's automation control capabilities. Because of the foregoing methods, the mining industry machinery's complexity in grabbing attention is also greatly reduced. In mechanical applications, it can also lower the likelihood of electrical mechanical failure. Furthermore, when the main themes are computer control, fault diagnostics, and sensor control covered in programmable control technology is used to control coal mining equipment and gear.

4.6.2 Power systems

Many auxiliary system applications, such as water treatment systems, are involved in the operation of electrical systems. These system actions must regulate and switch in a systematic manner to demonstrate the auxiliary control system's engagement in the application. During the development process, many industrial firms progressively altered their minds, insisting on applying the concept of scientific progress and sustainable development, as well as boosting the power system's economic benefits with the progressive growth of energy conservation and emission reduction work in recent years. Furthermore, in the power generation sector, we should strive for true energy savings and carbon reductions. At this moment, it is just as crucial to increase the auxiliary control level. PLC technology has been deployed as an auxiliary system by many significant organizations in China up to now. This allows for complete control of the entire manufacturing process, as well as the integration of the communication system into the PLC system. For instance, the floor of the main station and the distant IO station will be the most important components in the construction of the coal transport mechanism. The primary station layer is frequently designed in disease control instances. The two most critical components are the PLC system and the man-machine interface. In addition, an optical fiber will be used for the master station layer bus to link to a remote IO station. This has the potential to improve working conditions dramatically.

4.6.3 Air conditioning

Applying EAC is quite tough to break into the field of air conditioning. As a result, individuals must accept autonomous control technology as a given in order to avoid as many issues as possible in the control process. For the actual operation of the air conditioning system, on the other hand, there are numerous control systems. Better control effects should be represented in automatic air conditioner control. During application, this form of automatic control system can protect against outside undesired elements. It can customize program content based on user preferences and perform follow-up tasks. The use of PLC technology in the air conditioner electrical automated control system, for example, can enhance the cooling impact of central air conditioning. It can also make the air conditioning system more practical by guaranteeing that the system's multiple OS are compatible with the needs of the users

4.6.4 Traffic modeling

The electric automated control system is used in traffic, with traffic signal lights as the primary application object. Using this technology, logical traffic signal control can be made more evident, enhancing certain operations in the motor control system that is automated, and maintaining the control range expanding. When there is a traffic bottleneck on a specific road segment, for example, The electrical automatic control (EAC) system can collect information from the road according to the monitoring equipment's specific requirements and send the data to the appropriate management staff in a timely manner. The aforementioned behaviors are all examples of automated control content that do not demand the participation of relevant individuals. If this technology is fully integrated into the traffic EAC system, it not only lowers the risk of traffic collisions, but it also improves the visibility of each traffic route. This can also assist people in anticipating driving circumstances, making work and travel more convenient (Ghazal et al., 2016).

4.7 Design challenges and technical issues

4.7.1 Minimum degrees of security

Power lines are not always a safe medium to use. Attenuation of data is likely to be a problem on a power line network due to the presence of various elements. Residential appliances are expensive: A power line network modem's price is not always equivalent to a regular phone line network modem's price. A power line network modem's price is not always equivalent to a regular phone line network modem's price (Krishna et al., 2021).

4.7.2 Noise

Vacuum cleaners, light dimmers, and kitchen appliances, and drills are examples of noise sources that can interfere with the operation of a power line-based home network (Li, 2016).

4.7.3 Traffic management issues

In the case of high BER on a DPLC link, previously deployed commercial off-the-shelf routing equipment, designed to offer a network with a best-effort service, could not guarantee SCADA control traffic either priority or a minimum quality of service. As a result, excessive delays and data loss were experienced, greatly impeding routine gridline control operations.

4.7.4 Configuration issues

The system's setup (before to Marconi Selenia's involvement) necessitated a preliminary frequency adjustment of the DPLC devices (already in place). Specifically, the frequency range 88–200 kHz was exploited by using an 8 + 8 kHz radio frequency (RF) channel for each PLC link (operating mode being the single sideband mode with suppressed carrier). To adequately counteract channel attenuation, proper measurements of channel frequency response were conducted. Each PLC link was assigned a different RF channel due to high interference, which is a fairly regular occurrence on electric power lines, and precise tuning was required to obtain consistent transmission rates (Rao & Saraswathi, 2021).

4.8 Conclusions

PLC has a extensive history and heritage, as well as a wide range of applications for the Smart Grid implementations. However, the extensive use of PLCs could face various challenges to ensure secure communication and data privacy. For instance, PLCs extensively work to automate voltage and substations where it involves location, isolation, and fault restoration that necessitates low data rates for communication. This chapter provides a review of the role and status of PLCs. It depicts the overview of various PLC technologies and configurations of applications associated with different sectors. In this regard, some key concepts of PLCs for SG, including Distributed Grid, Advanced Metering Infrastructure, Vehicle to Grid Communication technologies, were considered and reviewed. Lastly, we provide a summary of various issues and challenges PLC technologies will face shortly. The future work discusses security requirements and technological advancements in the development of future SGs.

References

Al Zishan, A., Haji, M. M., & Ardakanian, O. (2021). Adaptive congestion control for electric vehicle charging in the Smart Grid. *IEEE Transactions on Smart Grid*, *12*(3), 2439–2449.

Alam, S., Sohail, M. F., Ghauri, S. A., Qureshi, I. M., & Aqdas, N. (2017). Cognitive radio based smart grid communication network. *Renewable and Sustainable Energy Reviews*, *72*, 535–548.

Al-Badi, A. H., Ahshan, R., Hosseinzadeh, N., Ghorbani, R., & Hossain, E. (2020). Survey of smart grid concepts and technological demonstrations worldwide emphasizing on the Oman perspective. *Applied System Innovation*, *3*(1), 5.

Avancini, D. B., Rodrigues, J. J., Martins, S. G., Rabêlo, R. A., Al-Muhtadi, J., & Solic, P. (2019). Energy meters evolution in smart grids: A review. *Journal of Cleaner Production, 217*, 702–715.

Borlase, S. (Ed.), (2017). *Smart grids: Infrastructure, technology, and solutions*. CRC press.

Chant, S. H., & McIlwaine, C. (2009). *Geographies of development in the 21st century: An introduction to the global South*. Edward Elgar Publishing.

Colak, I., Kabalci, E., Fulli, G., & Lazarou, S. (2015). A survey on the contributions of power electronics to smart grid systems. *Renewable and Sustainable Energy Reviews, 47*, 562–579.

Colak, I., Sagiroglu, S., Fulli, G., Yesilbudak, M., & Covrig, C. F. (2016). A survey on the critical issues in smart grid technologies. *Renewable and Sustainable Energy Reviews, 54*, 396–405.

Dell, R. M., & Rand, D. A. J. (2001). Energy storage—A key technology for global energy sustainability. *Journal of Power Sources, 100*(1-2), 2–17.

ETSI. www.etsi.org/WebSite/Technologies/Powerline.aspx.

Faheem, M., Shah, S. B. H., Butt, R. A., Raza, B., Anwar, M., Ashraf, M. W., … Gungor, V. C. (2018). Smart grid communication and information technologies in the perspective of Industry 4.0: Opportunities and challenges. *Computer Science Review, 30*, 1–30.

Fenton, D. R. (2006). *An experimental investigation into the electromagnetic compatibility aspects of high frequency power line communications* (Doctoral dissertation). The Open University.

Ferreira, H. C., Lampe, L., Newbury, J., & Swart, T. G. (Eds.), (2011). *Power line communications: Theory and applications for narrowband and broadband communications over power lines*. John Wiley & Sons.

Galli, S., & Logvinov, O. (2008). Recent developments in the standardization of power line communications within the IEEE. *IEEE Communications Magazine, 46*(7), 64–71.

Galli, S., Kurobe, A., & Ohura, M. (2009). *The inter-PHY protocol (IPP): A simple coexistence protocol for shared media. 2009 IEEE international symposium on power line communications and its applications* (pp. 194–200). IEEE.

Ghazal, B., ElKhatib, K., Chahine, K., & Kherfan, M. (2016). *Smart traffic light control system. 2016 Third international conference on electrical, electronics, computer engineering and their applications (EECEA)* (pp. 140–145). IEEE.

Huh, J. H. (Ed.), (2017a). *Smart grid test bed using OPNET and power line communication*. IGI Global.

Huh, J. H. (2017b). PLC-based design of monitoring system for ICT-integrated vertical fish farm. *Human-Centric Computing and Information Sciences, 7*(1), 1–19.

Huh, J. H., & Park, J. H. (2020). Decrepit building monitoring solution for zero energy building management using PLC and android application. *Sustainability, 12*(5), 1993.

Jin, C., & Kunz, T. (2011a). Smart home networking: Lessons from combining wireless and powerline networking. *Smart Grid and Renewable Energy, 2*(2), 136.

Jin, C., & Kunz, T. (2011b). *Smart home networking: Combining wireless and powerline networking. 2011 7th International wireless communications and mobile computing conference* (pp. 1276–1281). IEEE.

Kampker, A., Offermanns, C., Heimes, H., & Bi, P. (2021). Meta-analysis on the market development of electrified vehicles. *ATZ Worldwide, 123*(7), 58–63.

Kemal, M. S. (2019). Real-time monitoring of low voltage grids using adaptive smart meter data collection, Accessed online 23-07-2021.

Knapp, E. D., & Samani, R. (2013). *Applied cyber security and the smart grid: Implementing security controls into the modern power infrastructure*. Newnes.

Krishna, D. S., Rao, P. M., & Rao, T. S. (2021). *Emerging social information networks applications and architectures. Security in IoT social networks* (pp. 29–53). Academic Press.

Lambrechts, W., & Sinha, S. (2019). *Last mile internet access for emerging economies*. Springer International Publishing.

Li, L. (2016). *Frequent episode mining for smart home wireless sensor network* (Doctoral dissertation). ETH Zurich.

Liu, M., & McLoone, S. (2015). Enhanced AIMD-based decentralized residential charging of EVs. *Transactions of the Institute of Measurement and Control, 37*(7), 853–867.

Lorincz, J., Klarin, Z., & Ožegović, J. (2021). A comprehensive overview of TCP congestion control in 5G networks: Research challenges and future perspectives. *Sensors, 21*(13), 4510.

Mohassel, R. R., Fung, A., Mohammadi, F., & Raahemifar, K. (2014). A survey on advanced metering infrastructure. *International Journal of Electrical Power & Energy Systems, 63*, 473–484.

Muratori, M., Alexander, M., Arent, D., Bazilian, M., Dede, E. M., Farrell, J., ... Ward, J. (2021). The rise of electric vehicles—2020 status and future expectations. *Progress in Energy, 3*(2), 022002.

Mutoh, N., Ohno, M., & Inoue, T. (2006). A method for MPPT control while searching for parameters corresponding to weather conditions for PV generation systems. *IEEE Transactions on Industrial Electronics, 53*(4), 1055–1065.

Ndjiongue, A. R., & Ferreira, H. C. (2019). Power line communications (PLC) technology: More than 20 years of intense research. *Transactions on Emerging Telecommunications Technologies, 30*(7), e3575.

Oliveira, M. (2015). Country action plans in the ECOWAS. In *ECOWAS* sustainable energy policy & investment high level forum. Retrieved from www.ecreee.org/event/ecowas-sustainable-energy-policy-investment-high-level-forum.

Ourahou, M., Ayrir, W., Hassouni, B. E., & Haddi, A. (2020). Review on smart grid control and reliability in presence of renewable energies: Challenges and prospects. *Mathematics and Computers in Simulation, 167*, 19–31.

Ponce-Jara, M. A., Ruiz, E., Gil, R., Sancristóbal, E., Pérez-Molina, C., & Castro, M. (2017). Smart Grid: Assessment of the past and present in developed and developing countries. *Energy Strategy Reviews, 18*, 38–52.

Rao, P. M., & Saraswathi, P. (2021). *Evolving cloud security technologies for social networks. Security in IoT social networks* (pp. 179–203). Academic Press.

Roslan, M. F., Hannan, M. A., Ker, P. J., & Uddin, M. N. (2019). Microgrid control methods toward achieving sustainable energy management. *Applied Energy, 240*, 583–607.

Schmalz, S., Ludwig, C., & Webster, E. (2018). The power resources approach: Developments and challenges. *Global Labour Journal, 9*(2).

Schwarzenegger, A. (2007). Value of distribution automation applications.

Sharma, K., & Saini, L. M. (2017). Power-line communications for smart grid: Progress, challenges, opportunities and status. *Renewable and Sustainable Energy Reviews, 67*, 704–751.

Siano, P. (2014). Demand response and smart grids—A survey. *Renewable and Sustainable Energy Reviews, 30*, 461–478.

Throndsen, N. W., & Ryghaug, M. Smart system of renewable energy storage based on INtegrated EVs and bAtteries to empower mobile, Distributed and centralised Energy storage in the distribution grid.

Yigit, M., Gungor, V. C., Tuna, G., Rangoussi, M., & Fadel, E. (2014). Power line communication technologies for smart grid applications: A review of advances and challenges. *Computer Networks*, *70*, 366–383.

Trl, M. E., Trl, P. V., Trl, D. N., Damousis, Y., Theodoropoulos, T., Cirimele, V., & Doni, A. Task 3.3. 1 Review of existing power transfer solutions, D 3.3.1, 2014.

Zaglago, L., Chapman, C., & Shah, H. (2013, July 3–5). Barriers to nationwide adoption of the smart grid technology in Ghana. In *Proceedings of the world congress on engineering*. London.

Zhao, G., & Zheng, H. (2021). Analysis of practical application of PLC technology in electric automation. *Journal of Physics: Conference Series*, *1871*(1), 012017, IOP Publishing.

Zishan, A. A. (2020). On the control of electric vehicle charging in the Smart Grid.

Chapter 5

Key performance indicator-based energy management strategies for microgrid operation

P. Muralidhara Rao[1], P. Saraswathi[2] and Y. Vishnu Srinivasa Murthy[1]
[1]*School of Computer Science and Engineering, Vellore Institute of Technology, Vellore, India,*
[2]*Department of Computer Science and Engineering, GITAM Institute of Technology, Visakhapatnam, India*

5.1 Introduction

Microgrid (MG) data collection, automation, and control have been a research focus for the last three decades (Patnaik, Mishra, Bansal, & Jena, 2020). MGs have piqued the interest of scientists not only because of their potential to integrate grid integration of distributed generation in a more dependable and environmentally friendly way (with lower emissions) but also because of their high operational reliability and capacity in the event of natural calamities (Mahmoud & Azher Hussaina, 2014; Mohseni, Brent, & Burmester, 2019; Samad & Annaswamy, 2017). As a result, MGs play an active role in high operational reliability and capacity in natural calamities; however, there are still a few issues to work out regarding design, supervision, and the most efficient operation. For instance, Olivares et al. (2014) summarizes the combination of MGs into traditional electric systems and a description of the most important research, which includes DGs, energy converter applications, economic issues, operation and control, protection, and connectivity.

High operational reliability and capacity in the event of natural calamities in today's smart city practices and information and communication technology (ICT) is a critical component. The phrase "smart cities" was not coined in early 1980s literature on smart urban expansion. Rather, it arose in the late 2000s due to new technological breakthroughs such as digitization, the internet, the Internet of Things (IoT), artificial intelligence (AI), and the widespread use of smartphones. Many shards of evidence exist around the world

Sustainable Networks in Smart Grid. DOI: https://doi.org/10.1016/B978-0-323-85626-3.00003-X

that show the role of technological advancements in implementing social (inclusive) urban policy. Smart grid (SG) plays a vital role in the development of smart cities. Distributed systems are local RES-based generation units that have been deployed around the world, and integration challenges are arising (Elsayed, Mohamed, & Mohammed, 2015).

MGs offer a solution, in this case, allowing for purposeful islanding and connectivity to the public distribution grid based on the present status. MGs are composed of loads, generators, and storage units that may be controlled as a single entity. The optimization- and rule-based control techniques are investigated in this work for the cost-effective operation of grid-connected microgrids. A mixed-integer linear programming formulation is used with a model predictive control technique for optimization-based control. Renewable energy sources (RES) are becoming more popular due to green energy gaining popularity and lowering investment costs at a rapid rate around the world. Photovoltaic (PV) systems, in particular, are quickly being installed on distribution grids with lower voltages, causing grid integration challenges (Khare, Nema, & Baredar, 2013) that can be used to address these issues (Ibrahim, 2012). Control systems that include generation-based control, on the other hand, are unable to meet the owners of such DG units' supply stability needs.

Currently, the public distribution grid is in a state of disarray. A distribution grid-connected electrical MG is a grouping of loads, DG, and ESS that respond to the grid as a single programmable entity at the same coupling point. MGs can be used as building blocks for SG installation and deliver a large-scale DG grid integration solution (Aworo, 2019; Way, 2019). The capacity of MG operation to reduce local electricity costs is another compelling feature. Its control strategy and EMS are crucial in ensuring dependable, cost-effective, and secure operation (Marnay & Bailey, 2004; Prinsloo, Mammoli, & Dobson, 2017).

Of late, various power management strategies have been adopted for MG operation as rule-based control procedures, similar to (Zhu, Zhuo, & Shi, 2013). Optimization-based algorithms are used in modern control systems to ensure that the business runs smoothly while reaching the desired results to various control objectives simultaneously. Mohamed and Koivo (2007) presented multiobjective optimization for lowering a microgrid's operational expenses and emissions for each iteration. However, it does not guarantee peak performance over numerous time steps. Offline scheduling calculations over a temporal horizon are also popular (Morais, Kádár, Faria, Vale, & Khodr, 2010). The accuracy of load and RE production projections is critical to system functionality.

Several demonstration facilities that combine RESs and ESSs are operational worldwide and use the bulk hydrogen as a power storage medium, either in self-contained or grid-connected power generation systems (Gahleitner, 2013; Subburaj, Pushpakaran, & Bayne, 2015; Valverde, Ali, Abdel-Wahab, Guerra, & Hogg, 2013). The hydrogen storage technique is frequently combined with a battery bank for short-term energy storage in these facilities. Culminating in an MG, a hybrid polygeneration system bus stability is improved by properly integrating batteries, a hydrogen storage system, and transient load control because of the better flexibility and dependability given by combining multiple energy storage technologies.

One of the major issues facing today's society is balancing the multiple instances of the rising energy vs. falling energy dilemma, consumption and environmental protection. For example, intelligent energy grids, huge sensor networks collect data about the environment and react accordingly to optimize resource use. Thus SGs must first realize the current link concerning energy usage and a set of critical meteorological data. All "systems" that are intelligent (consumers, buildings, cities, and residences) can use their intelligence to adapt to changing climate circumstances (Hernández et al., 2012). In the end, this chapter provides a thematic and analytical perspective of KPIs based on energy management strategies for microgrid operations (Table 5.1).

TABLE 5.1 Acronyms used.

Acronym	Definition
SE	Smart environment
SG	Smart grid
RES	Renewable energy resources
DER	Distributed energy resources
DG	Distributed generation
SM	Smart meter
SC	Smart customers
SB	Smart building
EF	Emission factor
EMS	Energy management system
ESS	Energy storage systems
ICT	Information and communication technology
KPI	Key performance indicators
PV	Photovoltaic
PEL	Programmable electronic load
HITL	Hardware-in-the-loop
CERTS	Consortium for electric reliability technology solutions

5.2 Microgrid: an overview

The SG concept will be used to operate future electricity distribution systems. Advanced digital meters, distribution automation, communication technologies, and distributed energy resources are part of this intelligent system (Brown, 2008). Self-healing, high reliability and power quality, and the ability to accommodate a wide range of distributed generation and storage options are all desirable SG features (Wei, 2010). There have been a few modifications in the way the power systems function due to the constant increase in DER infiltration levels. ERs may comprise small-scale production (microsources) as well as large-scale production (macro sources). The microgrid under investigation mimics the behavior of a household photovoltaic system (Palaniappan et al., 2010). All aspects of the hydrogen chain are included in the microgrid: production (through an electrolyzer), storage, and energy generation.

The primary power distribution line also includes a battery bank. An electronic control power source models the behavior of a solar array in various weather situations. At the same time, an electrical load that can be programmed is used to replicate the demand for electricity in a home. During periods of abundant energy, an electrolyzer can be used to create hydrogen. After that, the metal hydride alloy storage tank holds hydrogen with a temperature control cooling/heating system. The hydrogen equipment setup is completed by a fuel cell. The bus voltage is maintained at 48 V with a battery bank of 24 enhanced lead-acid batteries. It is used to power all of the electronic equipment (Masa-Bote et al., 2014). The programmed power source emulates the PEL and duplicates the DC/AC converters connected to the bus where residential loads and exterior grids are connected. But the DC/DC converter that connects the photovoltaic array to the main 48-V bus performs poorly.

The DC system functions in grid-connected mode, which is achieved using HITL electronic emulation in both the electronic source of electricity and system and performance. By placing microsources near the load, transmission losses can be reduced. If there are major system disturbances, a microgrid can run in islanded mode by including renewable energy sources, regulated loads, and energy storage devices. As a result, the likelihood of a power outage at end-customers connected to a low voltage (LV) distribution grid can be minimized. Gas turbines, energy components, photovoltaic (PV) frameworks, wind turbines, and other microgrid components can provide power to clients. The majority of energy storage technologies use batteries and flywheels (Mohammed, Nayeem, & Kaviani, 2010).

In a microgrid, a storage device is required to maintain a balance between energy generation and consumption. Microgrids offer several benefits, including increased reliability, reduced emissions, improved power quality, supporting voltage, reduced voltage dips, and inexpensive energy supply costs. Utility companies can benefit from microgrids as well. The demand for distribution and transmission facilities can be decreased by utilizing distributed energy sources. They can reduce streams in transmission and dispersion circuits by identifying the right location near loads. Loss reduction and the ability to potentially substitute for network assets are the two most important effects of using DG. By locating a DG in the area of interest, the administration quality can improve the status of the end-users.

5.2.1 The power supply and programmed load devices

In particular, carry out the purchase and sell energy to the grid, respectively, which can mimic these processes because of the provision of a separate channel of output. The MG is outfitted similarly to the required sensors. Pressure regulates and monitors the hydrogen charging and discharging processes, transducers, and regulators ranging from 0 to 10 bar. In contrast, the temperature of the metal hydride tank's water cooling/heating system is monitored by type T thermocouples. To manage the hydrogen flow correctly, electrovalves are placed along the pipelines. A flow meter can monitor the tank's hydrogen flow in and out (Petrollese, Valverde, Cocco, Cau, & Guerra, 2016).

5.2.2 Traditional electric supply

A conventional electric supply system consists of generating plants that generate electricity, transmission cables for moving data, and electricity distribution lines that deliver energy over large distances to consumers and suppliers. Substations and transformation centers for transformers and transformer substations, which adjust the voltage to specific levels, are located between these elements. Fig. 5.1 depicts all of these components (Kok, Scheepers, & Kamphuis, 2010).

5.2.3 Advanced power supply models

With the introduction of the SG, a "smart" electric power distribution network that employs information and communication technology to optimize electric power generation and distribution systems is utilized. SGs enable more efficient power distribution based on energy demand. Simultaneously, several new DER, DG, and Dr control energy demand in real-time in response to market demands or the need to adjust the demand of current DG systems are all SG concepts that have emerged for future benefit (Fig. 5.2).

The SM is electric energy usage recorded using an electronic measurement system. The Smart Home (Björkskog et al., 2010) found that the amount of energy consumed increased at a quicker pace than population growth, indicating that energy consumption is an often disregarded part of

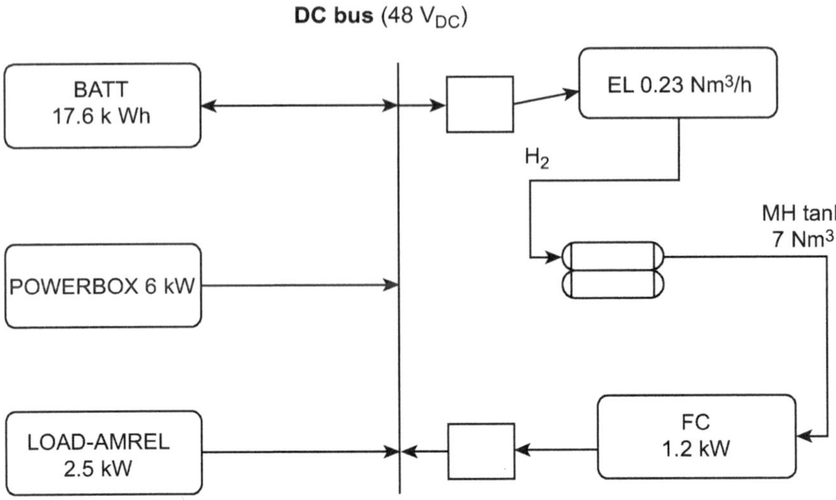

FIGURE 5.1 Configuration of the microgrid (Petrollese et al., 2016).

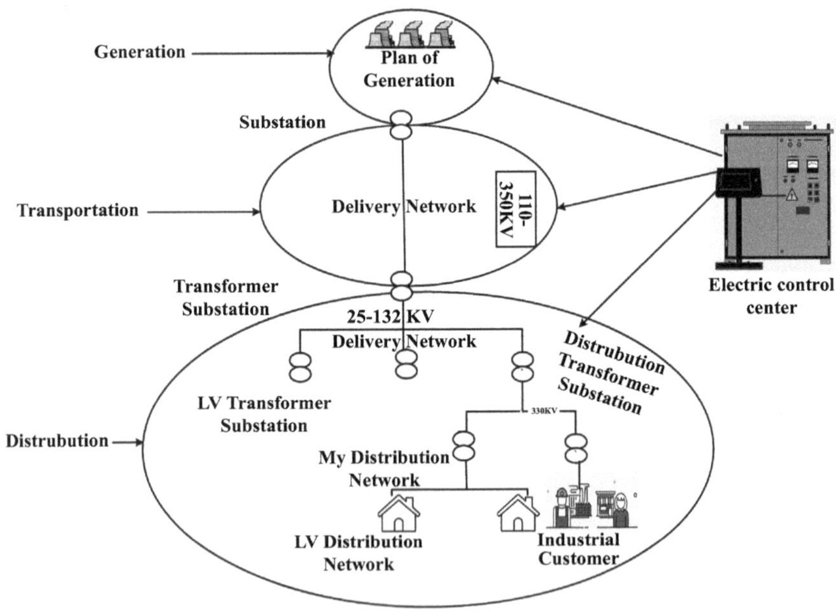

FIGURE 5.2 A generic system for electric power supply.

household life, and SC defines consumers who are conscious of what they consume and are examples of smart technologies. The European Technology Platform claims that SGs may expand their utilization by 2035 (Hernández et al., 2012).

5.2.4 Future technologies

SGs contribute to the formation of Smart Cities. Urban efficiency is increasingly dependent on the availability and efficiency of cognitive society and interpersonal infrastructure, in addition to the city's hard foundation (rational and social capital). ICT and sensors are critical in this context for identifying and meeting social and environmental requirements. A more specific notion emerges in this context: SB is a new type of building that includes RES with a system of intelligent management based on principles of economics or self-sufficiency (Carvalho, Costa, Marnoto, Sousa, & Vieira, 2018).

A SE can learn about its people and their physical surroundings and utilize that knowledge to improve the environment's performance. SE technologies can improve energy efficiency in various ways, including analyzing electricity use to spot trends and abnormalities, establishing a link between human behavior and electricity consumption, and simplifying surroundings to make them more energy-efficient.

Microgrids are a natural progression from SGs. A microgrid is defined as an "aggregate to deliver both electric power and heat," according to CERTS (Bibri & Krogstie, 2020), where "loads and micro-power units work together as a unified system." Power units, energy storage, and interconnected loads that can operate both connected to and independently from the bulk power system are required in the case of a power outage.

Energy supply security is critical for economic prosperity, automation, and modernization. Global energy demand is continuously increasing, and the world is currently concerned about meeting future energy demands. According to long-term estimates, global energy demand will significantly rise. Fossil fuels have been employed as key energy sources to meet this need. Fossil fuels produce greenhouse gases, which significantly impact the environment and future generations (Chen & Chen, 2011; Zhang, Fu, Geng, & Tao, 2011). In the meantime, the world's emission-free energy sources are renewable (solar, wind, hydro, geothermal, biomass, and so on). Renewable energy technologies are a great choice since they can contribute significantly to global power generation while producing fewer greenhouse gases (Department of Trade and Industry, U. K, 2003; Ugwuoke, Agwunobi, & Aliyu, 2012).

5.3 Energy consumption strategies and methodologies

5.3.1 Worldwide energy consumption: outline

High energy demand is caused by global population expansion and rising living standards. The rate of increase in global energy demand is outpacing the rate of population growth. Fossil fuels provide approximately 80% of all primary energy (Ummadisingu & Soni, 2011). The prediction of global energy use from 2002 to 2030 predicts an almost 60% rise in demand (1.6%/year).

By 2030, over 16.5 billion tons of oil equivalents (toe) will be consumed, up from 10.3 billion in 2002. Tables 5.2 and 5.3 show the global need for primary energy projections between 2002 and 2030 (Islam, Hasanuzzaman, Rahim, Nahar, & Hosenuzzaman, 2014) (Fig. 5.3).

In traditional AC power systems, an AC voltage source is converted to DC power and supplied to DC loads using an AC/DC inverter. Renewable resources are linked as distributed generators or AC microgrids as a result of the environmental difficulties associated with traditional power plants. When power can be supplied locally from renewable sources, long-distance high-voltage transmission is no longer essential. DC systems convert AC sources to DC and distribute the electricity via DC lines using power electronic converters. The efficiency of an AC or DC system is reduced as a result of these changes. A hybrid AC/DC microgrid was developed with the purpose of creating a more efficient system. Both the utilities and customers can gain from distribution energy. When clients use DGs, the quality and dependability of the power can improve (Elsayed et al., 2015). There will be fewer blackouts. The efficiency with which energy is used can be improved. The cost of electricity will be lower. Furthermore, greenhouse gas emissions can be greatly decreased. The system capacity will be raised and the decline will be minimized when utilities utilize DGs. It will be possible to improve the voltage

TABLE 5.2 The University of Seville microgrid's main components and characteristics.

Electronic power source		**Electrolyzer**	
Panel peak power	0.225 kW	H_2 net production rate	0.23 Nm^3/h@5 barg
Channel	2	Nominal power	1 kW
Manufacturer	POWERBOX	Manufacturer	HAMILTON-STD
Electronic load		**Fuel cell**	
Rated Power	2.5 kW	Nominal power	1.2 kW
Channel	2	H_2 rated consumption	18.5 Nl/min
Manufacturer	AMREL	Manufacturer	Ballard's NEXA
Batteries		**Hydrogen tank**	
Nominal voltage	12 V	Alloy	$LaNi_5$
Rated capacity (C_{120})	367 A h	Storage capacity	7 Nm^3
Manufacturer	EXIDE	Manufacturer	LABTECH

TABLE 5.3 Fossil fuel emission variables for electricity generation (Birol, 2007).

Fuel	EF (kg/kWh)		
	CO_2	SO_2	NO_2
Coal	1.2100	0.016	0.0048
Petroleum	0.8300	0.0182	0.0027
Gas	0.5500	0.0006	0.0008

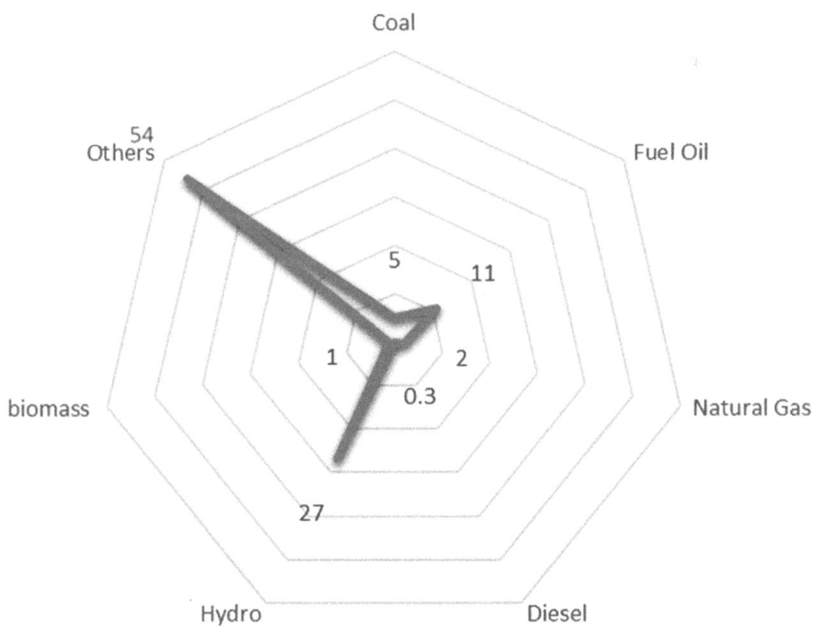

FIGURE 5.3 Malaysia's installed capacity share as of December 31, 2013 (Islam et al., 2014).

profile. It is possible to better manage reactive power. The relationship between the client and the utility will be improved (Ayres, Turton, & Casten, 2007).

According to the projections for worldwide primary energy consumption, the greatest amount of energy will come from fossil fuels, even if their proportion will marginally fall on or after 36% in 2002 to 35% in 2030 (Birol, 2007). The price of natural gas is high in the electricity industry, and its percentage change will have risen to a maximum of 60% *by* 2030, with a share of the market of 47% *by* 2030, up from 36% *in* 2002. Because of its great efficiency, most new power plants throughout the world are projected to use natural gas as a competitive fuel (British Petroleum, 2004; Goldemberg, Johansson, & Anderson, 2004). Coal's share of total primary energy demand will fall to 27% *by* 2035, down from 30% in 2012 (Global and Worldwide, 2011). Nuclear energy's share in the total energy mix will fall to 5% *by* 2030, down from 7% *in* 2002. Because of differences in economic growth, energy use by diverse users, energy pricing, location, economic arrangement, culture, lifestyle, and climate will all continue to vary in intensity in different regions (Bilen et al., 2008) (Table 5.4, Table 5.5, Table 5.6).

TABLE 5.4 Demand for primary energy in the world (Mtoc) (British Petroleum, 2004).

Sources of energy	2002	2010	2020	2030	2002–30
Coal	2384	2765	3190	3602	1.45%
Oil	3674	4304	5078	5762	1.55%
Gas	2195	2709	3453	4132	2.31%
Nuclear	694	774	778	767	0.45%
Hydro	222	278	323	367	1.75%
Biomass & waste	1120	1265	1426	1607	1.35%
Other renewables	60	102	164	254	5.65%

TABLE 5.5 Demand for primary energy in the world (percentage) from 2002 to 2030 (British Petroleum, 2004).

Sources of energy	2002	2010	2020	2030
Coal	24%	24%	23%	23%
Oil	35%	34%	34%	34%
Gas	20%	21%	23%	24%
Nuclear	6%	5%	5%	4%
Hydro	3%	3%	4%	3%
Biomass & waste	12%	11%	10%	11%
Other renewables	0.58%	0.86%	1.16%	1.62%

TABLE 5.6 Use of renewable energy around the world by kind and scenario (British Petroleum, 2004).

	Regulatory changes			Policies in place today		Scenario 450	
	2011	2020	2035	2020	2035	2020	2035
Electricity generation (TWh)	4484	7194	11,610	6846	10,020	7526	15,485
Bioenergy	422	760	1476	734	1250	796	2055
Hydro	3491	4552	5825	4410	5474	4667	6392
Wind	432	1324	2770	1197	2249	1440	4337
Geothermal	68	128	298	113	216	142	437
Solar PV	60	378	950	351	679	420	1386
Concentrated Solar power	2	42	246	35	121	56,804	56,804
Marine	1	2	40	2	25	2	65
Share of total generation	21%	25%	30%	26%	24%	27%	47%
Heat demand (Mtoc)	432	440	604	430	553	464	702
Industry	208	255	317	251	310	250	324
Building and agriculture	134	186	282	172	244	196	376
Share of total demand	9%	9%	13%	8%	12%	9%	17%
Biofuels (mboc/d)	1.2	2.0	4.2	1.8	3.4	2.7	7.5
Transportation via road	1.2	2.0	4.1	1.8	3.3	2.7	6.9
Aerospace	–	–	0.1	–	0.1	–	0.6
Percentage of total transportation	2%	5%	5%	4%	4%	6%	14%
Biomass (Mtoc)	746	728	682	731	688	719	646
Share of total bioenergy	58%	47%	38%	49%	41%	46%	30%
Demand for renewable energy as a percentage of overall demand	42%	34%	23%	35%	26%	31%	18%

According to the Paris-based International Energy Agency (IEA), energy consumption will increase by 60% by 2030 compared to 2001 levels (Khatib, 2012). It will be more difficult to meet energy demand by using fossil fuels as a major supply. It is possible to create power using fossil fuels. It has the potential to devastate the ecosystem and contribute to global warming. As a result, the future electricity grid should be carbon-free and environmentally favorable. Hydropower, solar, geothermal, wind, and biomass are all examples of renewable energy sources. These should be promoted as a primary energy source by policymakers. The prospect of the scope of today's renewable energy technology is enormous, and it has numerous advantages. RES-based power generation is on the rise, accounting for 20% of total electricity output in 2010 and anticipated to account for 31% by 2035. The OECD has a higher rate of renewable energy-based generation than other regions, as seen in Figs. 5.4 and 5.5 (about 18% in 2010 and 33% by 2035) (International Energy Agency, 2012).

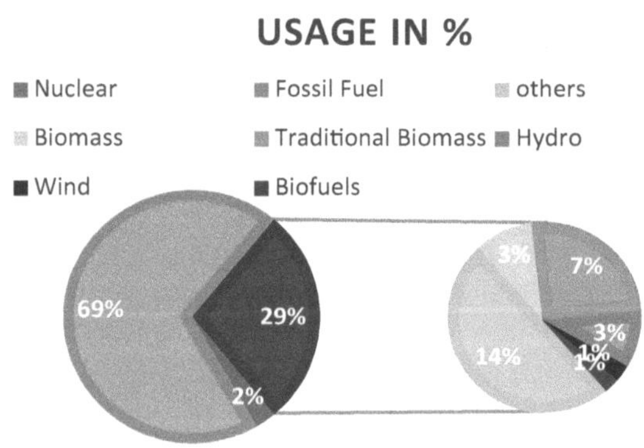

FIGURE 5.4 Global energy usage and renewable energy share, 2015 (Birol, 2007).

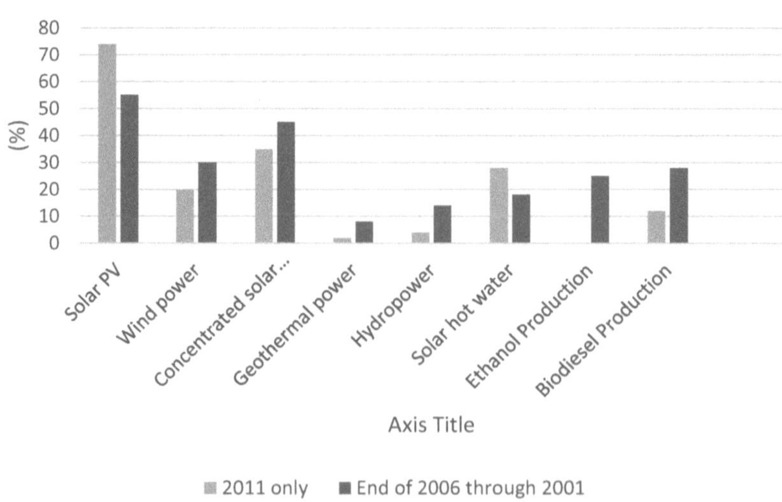

FIGURE 5.5 Rate of increase in renewable energy capacity from 2006 to 2011 (Birol, 2007).

5.4 Microgrid operations

According to Lopes, Moreira, and Madureira (2006), microgrids can operate in two modes. The first is grid-connected mode. The microgrid is connected to the upstream network in this mode. The main grid can provide all or part of the energy to the microgrid. Furthermore, when total production exceeds consumption, the power surplus can be transmitted to the main grid. The next mode is islanding, which occurs when the grid fails or when some planned actions, such as maintenance, require the microgrid to transition to islanded operation. As a result, the microgrid operates autonomously, which is referred to as island mode, in the same manner that real islands' electric power systems do.

5.4.1 Grid-connected mode

When controlled inverter interfaced microsources are operated, the PQ control method is used in connected grid mode. There are times when there are undulations in the loads and frequency and voltage disturbances, and the MG is managed by the main grid in this mode. The frequency and voltage of the system are not handled by microsources, and the microgrid simply adopts the frequency and voltage of the main grid. In power management systems, both active and reactive flow out of an inverter can be controlled in two ways: voltage-mode control and regulatory changes control. There are two types of power: active and reactive control, which both utilize the angle of phase and magnitude of the inverter AC voltage in voltage-mode control. In Park transform, below the *P*/*Q* control situation, the output voltage of the inverter is switched from abc to dq0. In addition, the voltage on the q axis is set to 0 ($u_{gq} = 0$) during the changeover. The inverters' output power is represented using the formulae below (Honarmand, 2015).

$$P_{ref} = u_{gd} \times i_{gd} + u_{gq} \times i_{gq} = u_{gd} \times i_{gd} \tag{5.1}$$

$$Q_{ref} = -u_{gd} \times i_{gq} + u_{gq} \times i_{gd} = -u_{gd} \times i_{gq} \tag{5.2}$$

i_g denotes the electricity of the grid and u_g denotes the voltage of the grid at the PCC

$$V_{sa}(t) = \nabla_s \cos(\omega_0 t \theta_0) \tag{5.3}$$

$$V_{sb}(t) = \nabla_s \cos(\omega_0 t \theta_0 - 2/3) \tag{5.4}$$

$$V_{sc}(t) = \nabla_s \cos(\omega_0 t \theta_0 - 4/3) \tag{5.5}$$

V_s is the peak phase-to-neutral value, ω_0 is the microgrid angular frequency, and θ_0 is the beginning power phase angle.

5.4.2 The island mode

The microgrid is no longer connected to the main grid, the DGs that were using the PQ control approach lose their voltage and frequency references while the microgrid is in islanding mode. The storage device that was previously controlled by a PQ control is now controlled by a V/f control. Before islanding mode, the storage device supplied the power difference between the main grid and the microgrid. The storage device is expected to have sufficient energy reserves to offer the same power difference as the main grid (Honarmand, 2015).

5.4.2.1 Smart grid: a security perspective

Security mechanisms should be upgraded from both a physical and cyber perspective to increase the grid's sturdiness and reliability. In the end, this measure will lessen the likelihood and effects of artificial disasters. Security of energy is a concept that ensures the availability of energy source reliability, ensures an adequate energy supply at a reasonable price, and prevents negative environmental consequences. Risk management, energy diversification, and policy implementation decision-making are all aspects of energy security. The use of SG technologies to integrate renewable energy can improve the electric system's energy security and safety (Sharifuddin, 2014) (Fig. 5.6).

A SG is a solution for modernizing the infrastructure and electrical energy system so that a more intelligent and dependable power grid can be presented. SGs have numerous advantages over traditional grids. SGs increase grid system dependability and sustainability by improving both physical and economic operations (Gangale, Mengolini, & Onyeji, 2013).

5.4.3 Electric microgrid

Microgrids have emerged as a feasible tool for accelerating the combination of DG units with the SG, lowering fossil fuel dependency, and increasing grid efficiency. The unbalanced nature of MGs could leverage their lack of inertia and poor energy storage capacity, as well as the large number and diversity of microsources, electronic power converters, and other circuits/devices. The vast quantity and diversity of microsources, as well as electronic power, all contribute to substantial parametric uncertainty and high failure rates. The basic architecture for an integrated MG into the electrical power supply is shown in Fig. 5.4. Energy storage, compensation devices, transformation devices, and linear and nonlinear loads in AC or DC are all examples of linear and nonlinear loads amongst the systems and subsystems described in MG (Ortiz, Gutiérrez, González, & Águila, 2020) (Fig. 5.7).

The distribution network operator (DNO), who is liable for the technical procedure of many MGs and is in command of the system's operational

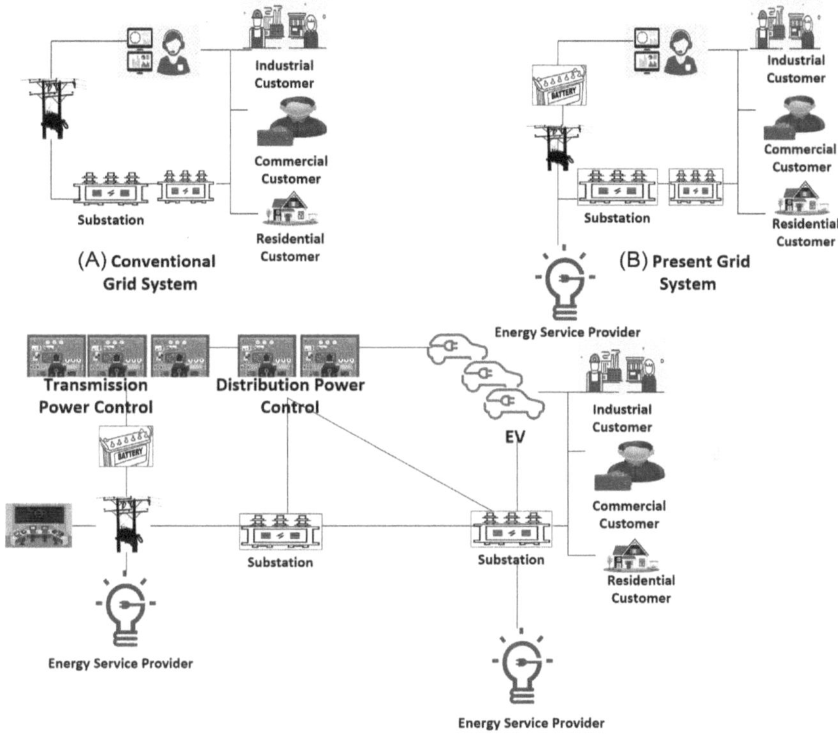

FIGURE 5.6 Systematic architecture for future grid.

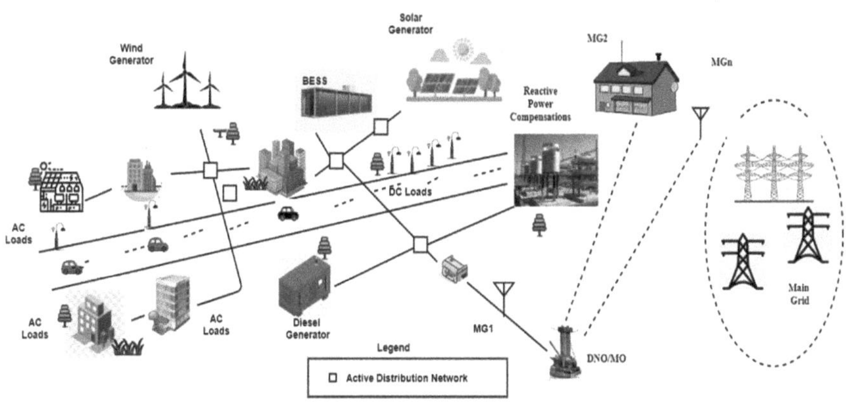

FIGURE 5.7 A typical architecture for electric microgrid.

functions and can integrate the MGs into the Main Grid (Jiayi, Chuanwen, & Rong, 2008). Instead, market functions are managed by the market operator (MO). To control voltage and frequency control while protecting the loads connected to the Main Grid and the various malfunctions and cyber-attacks that are revealed, MGs must be resilient. A justification for MGs as small-scale SGs can be found in numerous sources. Also consider that MGs offer an intriguing approach for improving energy flow in distribution networks, hence reducing energy losses. Without a doubt, MGs assist in the resolution of such issues and play a vital role in the new decentralized SG paradigm. MGs have a significant impact on the evolution of SGs in this regard, acting as suitable prototypes for solitary or connected sites.

5.4.4 Modeling and architectures of microgrid

The study of FTC techniques is critical to comprehending the dynamics of the HMG's multiple systems and subsystems, as well as to develop, analyze, and implement control systems, as well as to assess the repercussions of failures. Additionally, if needed, modeling approaches can be implemented. MGs, despite being small-scale electrical systems, have a significant level of complexity that changes throughout time as contrasted to standard energy systems (Gangale et al., 2013). Major obstacles are faced, especially when the MG is in fault mode (Saha & Johnson, 2018). Different modeling approaches are required because the topology and setup of models for MGs are dynamic and might change. The models must deal with DG in a dynamic manner with a short response time. As previously noted, MGs are unbalanced by design, have little energy storage capacity, lack inertia, a large number and variety of microsources, electronic power converters/other nonlinear circuits, significant parametric uncertainty, and high fault rates.

The different topologies utilized in MGs are summarized in Fig. 5.2. The MGs function in islanded mode, parallel to the main electrical grid (autonomous power), and in interconnected mode, as illustrated in the diagram. A hybrid microgrid that combines both AC and DC power sources is known as an AC/DC Hybrid Microgrid (HMG) (Fig. 5.8).

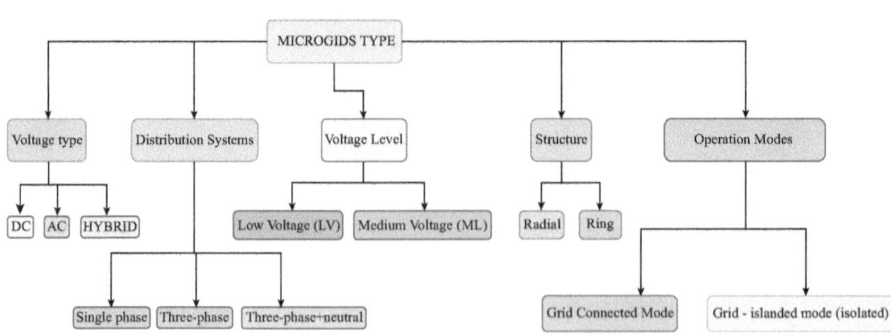

FIGURE 5.8 Typology.

By combining the primary benefits of AC and DC MG, this kind of MG provides a more suitable way (Elsayed et al., 2015; Mohammed et al., 2010). More research is needed on elements such as modeling fault tolerance to enable further integration of dispersed energy resources, design, and control mechanisms, as well as scalability, are required for HMGs to the Main Grid. Some researchers have attempted to create a linear time-invariant reduced order system (LTI) model using a time steady and a gain factor to simulate each source of DG, however this ignores grid dynamics. MGs are often represented by a DC source that is coupled to the primary bus and connected by a circuit breaker, a VSC, an RL filter, an elevator transformer, and a VSC. This method produces a low-order dynamic model that may be easily used for controller analysis and design.

5.5 KPI: Key performance indicators

This section will discuss the various key performance metrics that may be used to assess the microgrid. Power quality, dependability, economic indicators, and environmental indicators are among these metrics.

5.5.1 Power quality metrics

In this subsection, we will investigate power quality indicators as the first KPI. THD, demand, demand factor, and load factor are some of the subcategories of this category.

- *Power system harmonics*: This is not a new problem and was stated in (Grigsby, 2007). The waveform is in steady state when Fourier methods are employed to examine the distorted waveform. In fact, waveform distortion varies greatly depending on load levels and system conditions. Eq. 5.6 defines the THD in its simplest form.

$$\mathrm{THD}(\%) = \frac{\sqrt{\sum_{i=2}^{\infty} F_i^2}}{F_1} \times 100\% \tag{5.6}$$

The IEEE documents that cover IEEE limitations are IEEE 519–1992 and IEEE 519 A. According to these materials, there are two techniques to reduce harmonics in the power system. The first is a set of harmonic limits for the amount of harmonic current a user can input into the utility system at PCC. The harmonic voltage that the utility can provide to any customer at the PCC is the subject of the second set of harmonic limitations. Table 5.2 shows the IEEE 519–1992 voltage distortion limits. Acceptable distortion is minimized at higher voltage levels to reduce potential challenges for the majority of system users (Grigsby, 2007).

- *Demand*: According to (Grigsby, 2007), demand is a measurement of the energy required by loads over a short period of time, measured in terms of strength (kilowatts or Megawatts). Demand, often known as integrated demand in utility talks, can be measured across 10, 15, or 30 minutes. The formula below demonstrates how demand is calculated, where *EOI* denotes *Energy use over demand interval* and *DI* denotes *Dem Interval*

$$Demand = \frac{EOI}{DI} \tag{5.7}$$

- *Demand Factor*: As explained in (Grigsby, 2007), the demand factor is a ratio of the largest demand to the total connected load of a system or a component of a system under examination. The demand factor enables equipment in the power system to be properly monitored for normal loads, where *MD* denotes maximum demand and *Total CL* denotes total connected load.

$$Demand\ factor\ (DF) = \frac{MD}{Total\ CL} \tag{5.8}$$

- *The load factor*: It is similar to the demand factor in that it can be estimated based on energy consumption. The load factor is high if a load or collection of loads operates near its peak for the majority of the time. It is the ratio of energy use to demand multiplied by time (Grigsby, 2007).

$$Lod\ factor = \frac{Energy\ use}{Demand \times Time} \tag{5.9}$$

5.5.2 Reliability metrics

Reliability is the next KPI, which is divided into two categories: wind generation indices and power system reliability indices.

- *Wind generation indices*: The wind generation interrupted energy benefit (WGIEB) index, according to (Wang and Billinton, 2001), demonstrates the reliability of adding wind generating as a source of alternative energy.

$$\mathrm{WGIEB} = \frac{\mathrm{EEN}S_{\mathrm{bw}} - \mathrm{EENS}_{\mathrm{aw}}}{\mathrm{IWTGC}} \tag{5.10}$$

  where IWTGC denotes incrememtal WTG capacity, the energy could not be delivered after and prior to adding WTG units is represented by EENSbw and EENSaw, respectively. The cost savings from wind turbine outages (WGICB) index can also be used to determine the value of adding wind generating as a backup power source.

$$\mathrm{WGICB} = \frac{\mathrm{ECOST}_{\mathrm{bw}} - \mathrm{ECOST}_{\mathrm{aw}}}{\mathrm{IWTGC}} \tag{5.11}$$

where $ECOST_{bw}$ and $ECOST_{aw}$ are the projected system disruption costs after and prior to adding WTG units, respectively.

The index ENCG (equivalent number of conventional generating) is established to determine for given system, the number of WTG units necessary to replace a conventional generating unit of the same size, ECOST.

$$\text{ENCG} = \frac{\text{RNCG}}{\text{RNWTG}} \text{ for a specified ECOST} \quad (5.12)$$

The expected amount of CG and WTG is RNCG and RNWTG, respectively. The following is the definition of the equivalent conventional generating capacity (ECGC):

$$\text{ECGC} = \frac{\text{RCCG}}{\text{RCWTG}} \text{ for a specified ECOST} \quad (5.13)$$

where RCCG and RCWTG concern CG's and WTG's respective capacity requirements.

- *Power system reliability indices*: Currently, utilized reliability indices load point reliability indices and system reliability indexes are used to evaluate distribution system reliability was defined according to.

The following three aspects of load point reliability indexes:

1. λ : occurrences per year,
2. *r*: average duration of load interruptions (hours per occurrence),
3. *U*: average annual duration of load interruptions (hours per year).

5.6 Conclusion

Key performance indicators (KPI) are necessary to evaluate and optimize MG configurations. Moreover, these KPIs include economic indicators, reliability indicators, environmental indicators, power quality indicators and other related quality factors. Therefore this chapter emphasizes the available solutions that require the quantitative assessment of projects connected to microgrid development in isolated energy systems using representative KPIs, providing key stakeholders with a useful comparison of existing options. The evaluation method stresses the impact of operation methods and the values of the related control parameters on storage system sizing.

References

Aworo, O. (2019). *Resilient microgrids using a state controller* (Doctoral dissertation). University of Pittsburgh.

Ayres, R. U., Turton, H., & Casten, T. (2007). Energy efficiency, sustainability and economic growth. *Energy*, *32*(5), 634–648.

Bibri, S. E., & Krogstie, J. (2020). Environmentally data-driven smart sustainable cities: Applied innovative solutions for energy efficiency, pollution reduction, and urban metabolism. *Energy Informatics*, *3*(1), 1–59.

Bilen, K., Ozyurt, O., Bakırcı, K., Karslı, S., Erdogan, S., Yılmaz, M., & Comaklı, O. (2008). Energy production, consumption, and environmental pollution for sustainable development: A case study in Turkey. *Renewable and Sustainable Energy Reviews*, *12*(6), 1529–1561.

Birol, F. (2007). World energy prospects and challenges. *Asia-Pacific Review*, *14*(1), 1–12.

Björkskog, C., Jacucci, G., Mikkola, T., Bertoncini, M., Gamberini, L., Torstensson, C., ... Fiorentino, G. (2010). Beaware: A framework for residential services on energy awareness. In *International Conference on Mobile Ubiquitous Computing, Systems, Services and Technologies* (pp. 294–300).

British Petroleum. (2004). *The cost of generating electricity: A study carried out by BP power for the royal academy of engineering*. London, UK: The Royal Academy of Engineering.

Brown, R. E. (2008, July). Impact of smart grid on distribution system design. In *2008 IEEE Power and Energy Society General Meeting-Conversion and Delivery of Electrical Energy in the 21st Century* (pp. 1–4). IEEE.

Carvalho, J. M., Costa, R. V., Marnoto, S., Sousa, C. A., & Vieira, J. C. (2018). Toward a resource-based view of city quality: A new framework. *Growth and Change*, *49*(2), 266–285.

Chen, Z. M., & Chen, G. Q. (2011). An overview of energy consumption of the globalized world economy. *Energy Policy*, *39*(10), 5920–5928.

Department of Trade and Industry, U. K. (2003). *Energy white paper; our energy future: Creating a low carbon economy*. London: The Stationery Office.

Elsayed, A. T., Mohamed, A. A., & Mohammed, O. A. (2015). DC microgrids and distribution systems: An overview. *Electric power systems research*, *119*, 407–417.

Gahleitner, G. (2013). Hydrogen from renewable electricity: An international review of power-to-gas pilot plants for stationary applications. *International Journal of Hydrogen Energy*, *38* (5), 2039–2061.

Gangale, F., Mengolini, A., & Onyeji, I. (2013). Consumer engagement: An insight from smart grid projects in Europe. *Energy Policy*, *60*, 621–628.

Global, B. P., & Worldwide, B. P. (2011). *BP energy outlook 2030*. London, UK.

Goldemberg, J., Johansson, T. B., & Anderson, D. (2004). *World energy assessment: Overview: 2004 update*. United Nations Development Programme.

Grigsby, L. L. (2007). *Electric power generation, transmission, and distribution*.

Hernández, L., Baladrón, C., Aguiar, J. M., Calavia, L., Carro, B., Sánchez-Esguevillas, A., ... Gómez, J. (2012). A study of the relationship between weather variables and electric power demand inside a smart grid/smart world framework. *Sensors*, *12*(9), 11571–11591.

Honarmand, N. (2015). *Key performance indicators modeling for optimized microgrid configuration* (Doctoral dissertation).

Ibrahim, A. (2012). Renewable energy sources in the Egyptian electricity market: A review. *Renewable and Sustainable Energy Reviews*, *16*(1), 216–230.

International Energy Agency. (2012). *Iraq energy outlook: World energy outlook special report*. International Energy Agency.

Islam, M. A., Hasanuzzaman, M., Rahim, N. A., Nahar, A., & Hosenuzzaman, M. (2014). Global renewable energy-based electricity generation and smart grid system for energy security. *The Scientific World Journal*, 2014.

Jiayi, H., Chuanwen, J., & Rong, X. (2008). A review on distributed energy resources and MicroGrid. *Renewable and Sustainable Energy Reviews*, *12*(9), 2472–2483.

Khare, V., Nema, S., & Baredar, P. (2013). Status of solar wind renewable energy in India. *Renewable and Sustainable Energy Reviews*, *27*, 1–10.

Khatib, H. (2012). IEA world energy outlook 2011—A comment. *Energy Policy*, *48*, 737–743.

Kok, J. K., Scheepers, M. J. J., & Kamphuis, I. G. (2010). *Intelligence in electricity networks for embedding renewables and distributed generation. Intelligent infrastructures* (pp. 179–209). Dordrecht: Springer.

Lopes, J. P., Moreira, C. L., & Madureira, A. G. (2006). Defining control strategies for microgrids islanded operation. *IEEE Transactions on power systems*, *21*(2), 916–924.

Maams, M., & Azher Hussaina, S. (2014). Modeling and control of microgrid: An overview (Google Scholar).

Marnay, C., & Bailey, O. C. (2004). *The CERTS microgrid and the future of the macrogrid (No. LBNL-55281)*. Berkeley, CA (United States): Lawrence Berkeley National Lab (LBNL).

Masa-Bote, D., Castillo-Cagigal, M., Matallanas, E., Caamaño-Martín, E., Gutiérrez, A., Monasterio-Huelín, F., & Jiménez-Leube, J. (2014). Improving photovoltaics grid integration through short time forecasting and self-consumption. *Applied Energy*, *125*, 103–113.

Mohamed, F. A., & Koivo, H. N. (2007, April). Online management of microgrid with battery storage using multiobjective optimization. In *2007 International Conference on Power Engineering, Energy and Electrical Drives* (pp. 231–236). IEEE.

Mohammed, O. A., Nayeem, M. A., & Kaviani, A. K. (2010, July). A laboratory based microgrid and distributed generation infrastructure for studying connectivity issues to operational power systems. In *IEEE PES General Meeting* (pp. 1–6). IEEE.

Mohseni, S., Brent, A. C., & Burmester, D. (2019). A demand response-centred approach to the long-term equipment capacity planning of grid-independent micro-grids optimized by the moth-flame optimization algorithm. *Energy Conversion and Management*, *200*, 112105.

Morais, H., Kádár, P., Faria, P., Vale, Z. A., & Khodr, H. M. (2010). Optimal scheduling of a renewable micro-grid in an isolated load area using mixed-integer linear programming. *Renewable Energy*, *35*(1), 151–156.

Olivares, D. E., Mehrizi-Sani, A., Etemadi, A. H., Cañizares, C. A., Iravani, R., Kazerani, M., ... Hatziargyriou, N. D. (2014). Trends in microgrid control. *IEEE Transactions on smart grid*, *5*(4), 1905–1919.

Ortiz, L., Gutiérrez, L. B., González, J. W., & Águila, A. (2020). A novel strategy for dynamic identification in AC/DC microgrids based on ARX and Petri Nets. *Heliyon*, *6*(3), e03559.

Palaniappan, M., Gleick, P. H., Allen, L., Cohen, M. J., Christian-Smith, J., Smith, C., & Ross, N. (2010). *Clearing the waters: A focus on water quality solutions.*

Patnaik, B., Mishra, M., Bansal, R. C., & Jena, R. K. (2020). AC microgrid protection—A review: Current and future prospective. *Applied Energy*, *271*, 115210.

Petrollese, M., Valverde, L., Cocco, D., Cau, G., & Guerra, J. (2016). Real-time integration of optimal generation scheduling with MPC for the energy management of a renewable hydrogen-based microgrid. *Applied Energy*, *166*, 96–106.

Prinsloo, G., Mammoli, A., & Dobson, R. (2017). Customer domain supply and load coordination: A case for smart villages and transactive control in rural off-grid microgrids. *Energy*, *135*, 430–441.

Saha, S. S., & Johnson, N. G. (2018, August). Point-on-wave analysis of three-phase induction motor drive under fault external to the power plant. In *2018 IEEE Power & Energy Society General Meeting (PESGM)* (pp. 1–5). IEEE.

Samad, T., & Annaswamy, A. M. (2017). Controls for smart grids: Architectures and applications. *Proceedings of the IEEE*, *105*(11), 2244–2261.

Sharifuddin, S. (2014). Methodology for quantitatively assessing the energy security of Malaysia and other southeast Asian countries. *Energy Policy*, *65*, 574–582.

Subburaj, A. S., Pushpakaran, B. N., & Bayne, S. B. (2015). Overview of grid connected renewable energy based battery projects in USA. *Renewable and Sustainable Energy Reviews, 45*, 219–234.

Ugwuoke, P. E., Agwunobi, U. C., & Aliyu, A. O. (2012). Renewable energy as a climate change mitigation strategy in Nigeria. *International Journal of Environmental Sciences, 3* (1), 11–19.

Ummadisingu, A., & Soni, M. S. (2011). Concentrating solar power–technology, potential and policy in India. *Renewable and sustainable energy reviews, 15*(9), 5169–5175.

Valverde, L., Ali, D., Abdel-Wahab, M., Guerra, J., & Hogg, D. F. (2013, May). A technical evaluation of Wind-Hydrogen (WH) demonstration projects in Europe. In *4th International Conference on Power Engineering, Energy and Electrical Drives* (pp. 1098–1104). IEEE.

Way, B. (2019). *Improving the intersect of the power distribution system and the built environment in developing countries* (Doctoral dissertation). Heriot-Watt University.

Wei, C. (2010, March). A conceptual framework for smart grid. In *2010 Asia-Pacific Power and Energy Engineering Conference* (pp. 1–4). IEEE.

Wang, P., & Billinton, R. (2001). Reliability benefit analysis of adding WTG to a distribution system. *IEEE Transactions on Energy Conversion, 16*(2), 134–139.

Zhang, J., Fu, M., Geng, Y., & Tao, J. (2011). Energy saving and emission reduction: A project of coal-resource integration in Shanxi Province, China. *Energy Policy, 39*(6), 3029–3032.

Zhu, Y., Zhuo, F., & Shi, H. (2013, June). Power management strategy research for a photovoltaic-hybrid energy storage system. In *2013 IEEE ECCE Asia Downunder* (pp. 842–848). IEEE.

Chapter 6

Smart energy optimization using new genetic algorithms in Smart Grids with the integration of renewable energy sources

Reddi Sivaranjani[1] and P. Muralidhara Rao[2]
[1]*Department of Computer Science and Engineering, Anil Neerukonda Institute of Technology and Sciences, Visakhapatnam, India,* [2]*School of Computer Science and Engineering, Vellore Institute of Technology, Vellore, India*

6.1 Introduction

In general, typical smart grids (SGs) follow the central management principle (Khalid et al., 2018) to ensure control operation of all the units connected to it at the highest efficiency. In addition to that, it assists energy management by reducing electrical consumption during peak energy utilization times.

The SG was first introduced in Italy. Because of its effectiveness, many countries are using SG technology. This technology has been extensively adopted and employed for its massive benefits, like Demand Side Management (DSM) (Jindal, Singh, & Kumar, 2018; Sarkar, Chakrabarti, & Bhattacharyya, 2020), decreased greenhouse gas production, shared energy in an effective manner, stored electricity, and data for analysis and motivation in making intelligent decision making. In a typical SG, the user connects smart appliances via data network, the smart meter collects the data from appliances, which is then directed to the smart microgrid. The components of a SG consist of:

1. Smart Meter: Records the amount of energy used by smart appliances at different intermissions, example multiples of fifteen minutes. Also, directs data recorded for further processing.
2. Demand Management: Manages the electric appliance of various users. Further divides into DSM and Demand Response (Dr) (Good, 2018; Sanjeev, Padhy, & Agarwal, 2018). Its objective is to manage electricity

Sustainable Networks in Smart Grid. DOI: https://doi.org/10.1016/B978-0-323-85626-3.00006-5

loads and lessen the operative price to the customer. In Dr, customers regulate their electrical load accordingly.

3. Prosumer: An individual who consumes electricity as well as produces energy by integrating their own renewable energy. They can either store energy for future use or share to other consumers. They offer a chance to manage energy at minimum cost and inclines to financial and ecological profits. The challenges are:
   a. Minimizing dependency on the main grid (Alam, St-Hilaire, & Kunz, 2016; Rathnayaka, Potdar, Dillon, & Kuruppu, 2015)
   b. Effective procedure to share energy
   c. Identification of best cost estimation policy (Barbato & Capone, 2014; Khan, Mahmood, Safdar, Khan, & Khan, 2016; Tan, Ramachandaramurthy, & Yong, 2016)
   d. Effective appliance scheduling approaches to accomplish customer favorites.

6.2 Structure of modern Smart Grid system

The structure of SG, in the view of electricity circuit defines complete arrangement of devices, three main components were present in the arrangement namely traditional energy producers, energy producers using renewable energy sources and consumers.

- Traditional energy generators: Thermal plants, hydroelectric, and nuclear power stations are the major energy producers of domestic and industrial consumers.
- Renewable energy producers: Alternative energy generators are in the form of renewable energy sources (also called micro producers) having limited capacity of electricity generation from the renewable resources such as PVs and horizontal and vertical windmills.
- Consumers: Consumers are units that utilize electricity coming from both traditional and renewable sources. In general, consumers are categorized into domestic, industrial, and commercial consumers. The domestic consumers have several electricity consumption units like all the electrical appliances present in house. In general, appliances are of four types namely,
  - Non-interruptible Appliances (NIA): The appliances which works uninterruptedly once started continues till the completion of its cycles. Ex: washing machine, cloth dryer, microwave oven
  - Schedulable Appliances (SA): The appliances can be scheduled at customers convenience time with in given horizon. Ex: dish washer, iron, heat pump, TV, laptop, computer, Electric vehicle and vacuum cleaner.
  - User dependent Appliances (UDA): The appliances, which can be operated according to the user wish or presence. Ex: lights, and fans.
  - Temperature dependent (TD): The appliances work in accordance with the temperature. Ex: Air-Conditioner, Freezer, refrigerator etc.

The appliance will be in ON when temperature is low than specific level, otherwise in OFF mode.

Let Ap = {NIA, SA, UDA, TDA} be appliances set, for $a \in A_p$, a be an appliance and P_a be energy consumption vector of "a":

$$P_a = [P_a^1, P_a^2, P_a^3, \ldots, P_a^{24}] \tag{6.1}$$

Overall power consumption in a day can be written as:

$$p_t^{T_h} = \sum_{h=1}^{24} x_a^{(h)} \tag{6.2}$$

where T_h is total time taken in a day, written as

$$T_h = [1, 2, \ldots, 24] \tag{6.3}$$

Our objective is to optimize P_a. Therefore, overall energy consumed by A_p in a day as:

$$E_T = \sum_{t=1}^{|T_h|} \sum_{a}^{|\mathrm{Ap}|} (x_{t,a}) \tag{6.4}$$

The detailed information related to power consumption of house hold appliances (Ayan & Turkay, 2018) is described in Table 6.1.

TABLE 6.1 Appliances and energy usage.

Appliance	Priority	Power (W)	Start time (h)	End time (h)
Refrigerator	1	550	1	24
TV	1	150	18	23
Oven	1	2000	18	20
Desktop	1	150	21	22
Freezer	1	200	20	24
Laptop	1	100	1	16
Air conditioner	2	1140	14	22
Vacuum cleaner	2	600	19	23
Dish washer	3	1300	19	21
Washing machine	3	1400	20	22
Iron	3	2000	21	22
Tube lights	1	30	5	19
Fans	1	30	5	1

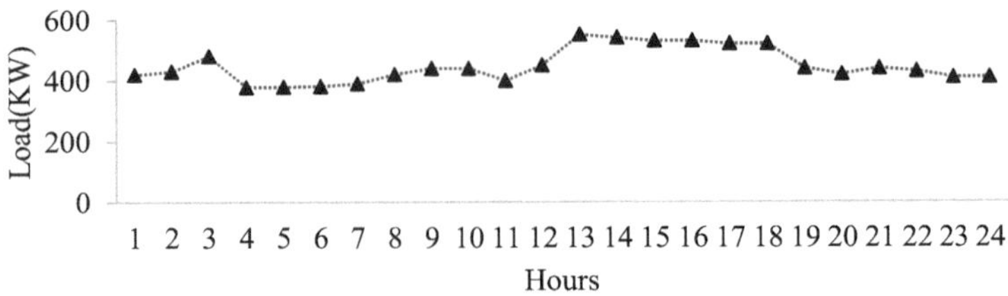

FIGURE 6.1 Load distribution on the smart meter for 24 h.

Fig. 6.1 shows the load distribution on the smart meter for 24 hours depending upon the energy consumed by each appliance (Mary, 2014), we can observe that load on smart meter is increasing from the 10:00 a.m. onwards till evening (5:00 p.m.) as the majority of the appliances in use, because of the usage of the fans, tv, lights in the evening results lesser load on smart meter.

6.3 Renewable energy integration

Energy is the major requirement to a smart home; without energy, no activity is possible. All the electronic appliances in a smart house require energy. According to its procedure involved in electricity generation, it is mainly divided into two categories: renewable and non-renewable energy.

1. Renewable energy: also referred as natural energy, usually generated from natural sources or process. Examples: solar, wind, hydro, tidal energy, geothermal energy, bio fuel, biomass, ocean thermal, biogas energy etc.
2. Non-renewable energy: It refers to the natural source found beneath the earth and is a limited resource. These resources do not replenish at the same speed at which it is used up, usually takes millions of years to develop. Examples: fossil fuels, geothermal energy and nuclear energy are part of this group.

6.3.1 Introduction to renewable energy source

Renewable energy sources get their energy from natural energy flows like sunlight, wind, flowing water, biological processes, and geothermal heat flows. At present, the most promising alternative energy sources include wind, solar, hydroelectric, geothermal and ocean energies, biomass and ethanol etc. The rapid development and advancement in technological of distributed energy resources have arisen a thought of integrating different types of renewable energy sources, for example, PVs and wind turbine, with lower capacities in the range of kW to MW. In this chapter, solar energy is used as the RES.

6.3.1.1 Solar energy

Modern housing uses solar power systems as the RES, which generally use PV to gather the sun's radiation. PV cells use radiation to create electricity. Each cell comprises a semiconductor material, silicon is the most commonly used material in one of several forms with impurities diffused all over, and enveloped with silk screen. Cells are combined in the form of circuit and mount them like a module. The semiconductor material let the electrons to be discharged from impurities by sunrays to move fast into the circuit, makes power. In general, PV modules integrated in residential areas produces power is approximately more than 10 W and lesser than 300 W. An inverter must present in PV is useful to convert DC into AC in order to use it by electrical appliances and compatible with electric grid, also used in establishment of large power plants. The generated electricity from these modules can significantly diminish pollution. Advancement in PV technology is more effective, inexpensive, standardized, consistent and long serving modules will be available. PV systems' value to the energy sector specifically in suburban capacities, is progressively apparent. However, high costs in PV system installation scares homeowners from purchasing. But, encouragement in installation of PVs system can be done by motivating consumers to buy such systems. Further, to encourage consumers in purchase of PV systems and evolution in integration of RES at a faster rate, the administration should provide an inspiration package to help homeowners towards PV systems installation, especially on new buildings.

6.3.2 Integration of renewable energy resource

Advancements in the renewable generation (Baronti, Vazquez, & Chow, 2018; Erdinc, Tascikaraoglu, Paterakis, Eren, & Catalao, 2017; Luo, Huang, & Gupta, 2018) method makers have been empowered to expand the supportability of the electric power framework. The proposed architecture has integrated solar PV system as the RES and the solar Borowy's energy model (Abushnaf & Rassau, 2018). Solar PV technology is being used; it converts solar energy into electricity. PV effect refers to the process of converting photons from sunshine into electricity (voltage) using semiconductor materials. Borowy suggested a simple approach for determining the output energy of a PV module. This model uses Maximum Power Point Tracking, it's an arrangement, where the Inverters connected to the grid are used to get the most power out of a PV board. The released output power of a PV module is computed in this model by multiplying voltage (V) and current (I). The power outputs in Borowy's model are determined by sun brightness and PV module parameters. As a result, PV module attributes determine our desired output, which is verified by

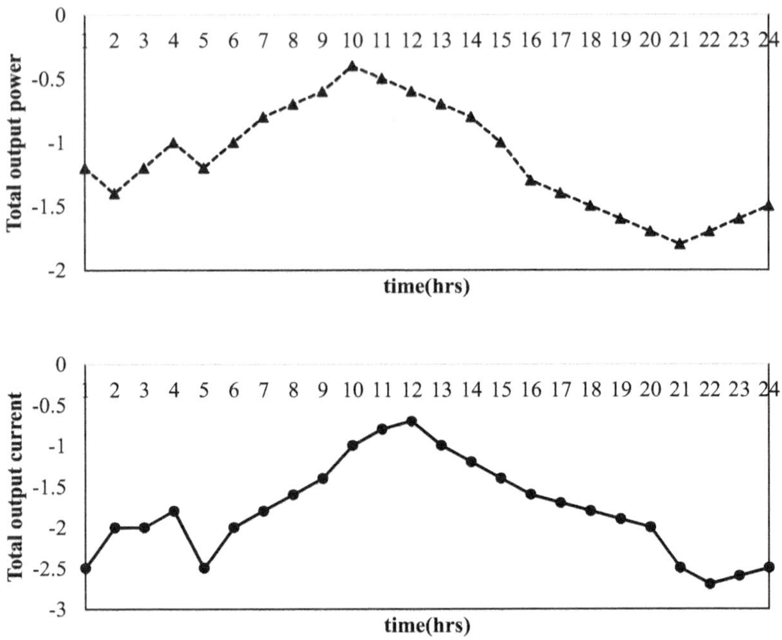

FIGURE 6.2 Borowy's model based total power and current.

module capacity. The overall power produced is 99.46×10^{17}. Fig. 6.2 shows the overall power and current generated (Zhao, Lee, Shin, & Song, 2013). In general, the maximum yield power of the PV array on day and at hour is calculated using PV characteristics under certain conditions (cell temperature at 25°C and the solar radiance is 1 kW/m^2), ambient temperature and irradiation conditions. To estimate the voltage V_{out} and I_{out} current, it is divided into parts that are described below:

Changes in temperature (ΔT) described as:

$$\Delta T = T_{am} + \left(\frac{\text{NCOT} - 20^{\circ}C}{800}\right)\text{G} - 25 \tag{6.5}$$

where T_{am}, NCOT represents ambient and nominal cell operating temperature and G is solar irradiance incident on surface of PV module placed at angle $\beta\left(\frac{W}{m^2}\right)$.

Change in current (ΔI) can be subject to solar irradiance and temperature, difference can be determine by:

$$\Delta I = K_1\left(\frac{G}{1000}\right)T_{am} + \left(\frac{\text{NCOT} - 20^{\circ}\text{C}}{800}\right)\text{G} - 25 + \left(\frac{G}{1000} - 1\right)G\frac{I_{SC}(\text{CEM})}{1000} \tag{6.6}$$

K_1 is temperature coefficient (A/°C) and I_{SC}(CEM) is PV module short circuit current under STC (A)

Change in voltage (ΔV) is subject to ΔT, ΔI, and resistance, its value is determined by:

$$\Delta V = -K_V \Delta T - R_s \Delta I \tag{6.7}$$

where K_V is open-circuit voltage temperature coefficient (V/°C)

PV module output current is calculated as:

$$I = I_{SC}\left[1 - C_1\left[\exp\left(\frac{V + \Delta V}{C_2 V_{oc}} - 1\right)\right]\right] + \Delta I \tag{6.8}$$

where

$$I_{SC} = G\frac{I_{SC}(\text{CEM})}{1000},$$

$$C_1 = \left(1 - \frac{I_{mpp}}{I_{SC}}\right)\exp\left(\frac{-V_{mpp}}{C_2 V_{oc}}\right),$$

$$C_2 = \frac{\frac{V_{mpp}}{V_{oc} - 1}}{\ln\left(1 - \frac{I_{mpp}}{I_{SC}}\right)}$$

where V_{oc} is PV module open-circuit voltage (V),

Maximum output power is determined by:

$$P_{max}^{out} = VI \tag{6.9}$$

6.3.3 Smart Grid electricity pricing policies

Pricing is the key components in smart grid besides billing options, many varieties of tariffs offer more choices allows the customer to identify best fit tariff. The venders offer a united electricity bill by considering energy, network costs, taxes and additional charges. They also offer a pricing list with a variety of options that allows people to choose the best price based on their potential and risk aversion. Fixed price offers or various dynamic pricing offers are examples of such deals. For a set length of time, fixed pricing guarantees the same price for energy components irrespective of change in market price that is, consumers get the same prices for all hours during the defined period even tough price variations is present.

In reality, fixed-price contracts grew increasingly popular among consumers. The alternative arrangement to the fixed electricity pricing is smart pricing, where retailers provide dynamic electricity prices at different times. Depending on the process, consumers naturally prefer to use less electricity when the electricity is more, and results decrease in demand during the peak hours. In other words, customer electricity usage will dynamically change the electricity price, instead of directly controlling their loads. In general, dynamic electricity pricing techniques are classified into:

- Time-of-Use (TOU) (Bui, Hussain, & Kim, 2018):The price per kWh depends on the time, simple day and night prices, or, for example, on-peak and off-peak hours, which divide the day into multiple quiet intervals.
- Critical Peak Pricing (CPP) is a top-up rate, electricity prices significantly increase for certain days in a year when wholesale charges are peak.
- RTP (Hongbo, Yan, & Yong, 2018) wholesale electricity prices are calculated based on at least hourly metering of consumption, or with even higher granularity (e.g., 15 minutes).
- IBR (Asgher, Rasheed, & Awais, 2018): It's an electricity pricing mechanism adapted by many power utilities since 1980, this price is calculated at two levels termed as lower and higher, consumers who uses more power the more electricity bill to be paid per kWh. Especially, the electricity bill value will climb to higher value when the user's hourly or regular or once-a-month electricity usage beats particular limits. So, the customers opting for this tariff usually distribute their energy usage among distinct timings in a day to evade higher charges also in reducing of grid's PAR.

The dynamic electricity charge tariffs of all four categories are shown in Fig. 6.3.

From the graphs an RTP dynamic electricity pricing is more effective than IBR, TOU and CPP. In our chapter, IBR pricing is adapted and is being combined with RTP. Anyway, paying attention is essential on real-time appliances usage at low peak hour. In our technique, an integrated pricing model is used, RTP and IBR would offer price variations on utilized power.

Consider that a customer desires to minimize their electricity bill and prepare themself to use more appliances in the morning around 5:00 a.m. However, the energy charges in morning time are relatively high due to heavy load in demand. So, using appliances earlier may add an extra power bill, due to change in power costs. We adopted the RTP + IBR method for electricity bill pricing (EP) expressed as:

$$p_{\text{tch}} = \begin{cases} a^h, \text{if} 0 \le t_{\text{pcm}} \le \text{th}_1 \\ b^h, \text{if} \text{th}_1 < t_{\text{pcm}} \le \text{th}_2 \\ c^h, \text{if} t_{\text{pcm}} > \text{th}_2 \end{cases} \tag{6.10}$$

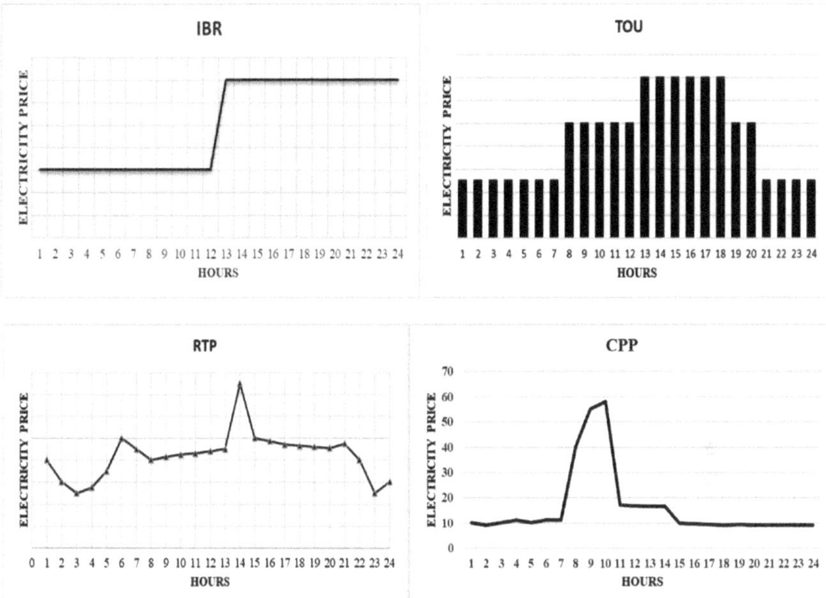

FIGURE 6.3 Dynamic electricity price tariffs inclining block rate, TOU, real time energy price and critical peak pricing. *TOU*, Time-of-Use.

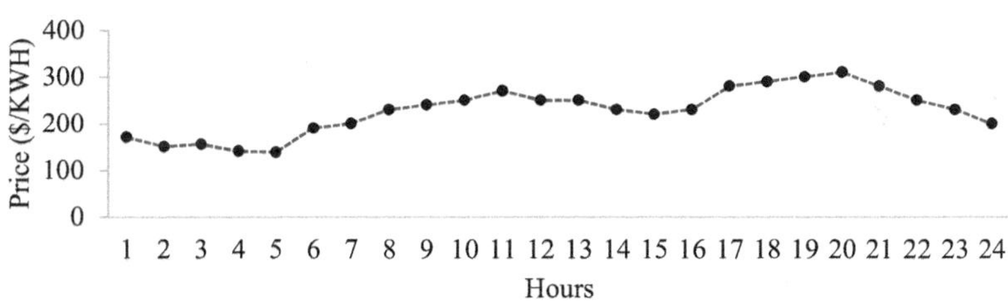

FIGURE 6.4 Average hourly pricing graph.

The EP used in the proposed work maintains two threshold levels (th_1 and th_2) while calculating electricity bill, when $t_{pcm} \leq th_1$, then EP is a^h level/$/kWh. Otherwise, it would be in b^h level/$/KWh whenever the power consumption is more than th_1 and lesser than th_2 or it will be in $c^h/\$/KWh$ when the power consumption is beyond the threshold level th_2. In order to diminish cost and PAR, smart home have an additional arrangement of RES, which is used to run home appliances. Fig. 6.4 shows an hourly average price for 24 hours collected from the California Independent System Operator CAISO (Dahlke, 2020).

6.3.4 Problem formulation

The aim is to find the optimal scheduling arrangement within the agreed limits and the favorites. Also taking into consideration that residents have certain intervals to set up their load prior to scheduling. Suppose "*a*" number of loads scheduled in T_h hours (equally spaced 24 hours) such that P_a should encounter the energy in demand. However, $P_a^{\min}$ and $P_a^{\max}$ represent lower and upper power limits, so s_a and e_a represents starting and ending time limits. Meanwhile, lt_a shows no. of time slots of "*a*." Also, t_a be "*a*'s" operation time is shown in Fig. 6.5.

For each $a \in A_p$, a group of parameters termed OTI $\in [s_a, e_a]$, LOT lt_a, power consumed value per hour x_a and OST t_a. Having s_a, e_a and lt_a, t_a should be $\geq s_a$, and $\leq e_a - s_a$. In particular, t_a can be written as:

$$t_a \in [s_a, e_a] \tag{6.11}$$

The power usage of "*a*" is 0 when t_a operating timings are beyond the starting and ending time limits. Otherwise, P_a represented as:

$$P_a^{\min} \leq P_a \leq P_a^{\max} \tag{6.12}$$

Now, P_{sch}
Determined as:

$$P_{\mathrm{sch}} = \sum_{t=1}^{24} \sum_{a}^{|A_p|} x_{t,a} \tag{6.13}$$

Fig. 6.6 shows factors used to represent delay actor (DF), which separates the incidence of two events. However, we are aiming to diminish the delay

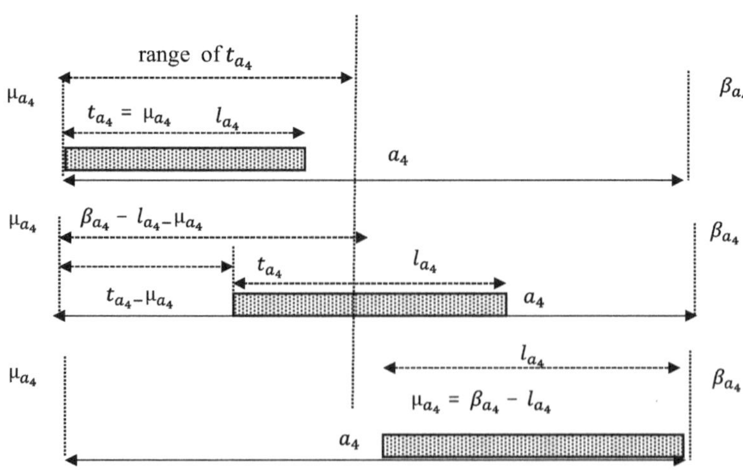

FIGURE 6.5 Impact of delay factor.

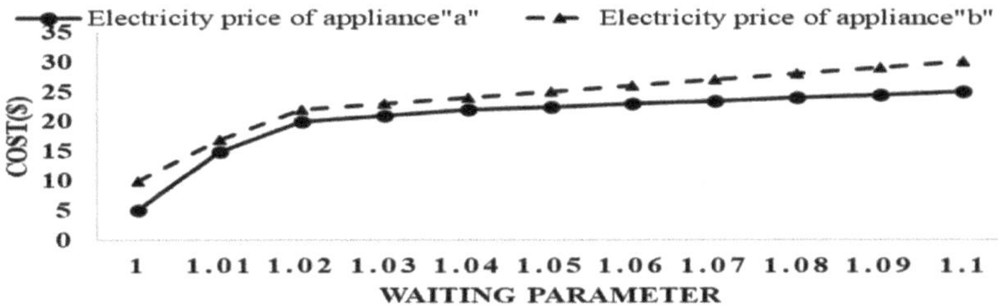

FIGURE 6.6 Impact of ζ on electricity bill.

time to increase user comfort. Here, DF of "a" is explained based on (Javaid et al., 2017; Rasheed, Javaid, Ahmad, Awais, et al., 2016; Yaagoubi & Mouftah, 2015; Zhao, Lee, Shin, & Song, 2013)

$$\mathrm{DF}_a = \frac{t_a - s_a}{e_a - \mathrm{lt}_a - s_a} \tag{6.14}$$

The max and min limits of the DF are fixed to 0 and 1. Though, if any appliance $t_a = s_a$ then DF_a become 0. If t_a of any appliance is $e_a - \mathrm{lt}_a$ then DF_a would be 1. It also, shows the relationship where $\mathrm{DF}_a = 0$ and $\mathrm{DF}_a = 1$. Now, a delay parameter ρ and the appropriate formula is $\sum_{a \in A_p} \rho^{\mathrm{DF}_a}$. The sum of appliances DFs by including waiting time cost expressed as:

$$\rho^{\mathrm{DF}_a} = \frac{\zeta_a^{e_a - h}}{p_t^{T_h}}, \tag{6.15}$$

for each a $\in A_p$, higher the ζ_a(a'scontrol parameter) value, higher the cost that is,

$$\sum_{t=1}^{24} \sum_{a}^{|A_p|} \rho^{\mathrm{DF}_a} x_a^h \tag{6.16}$$

Here, the $\rho \geq 0$ for every "a" in $h \in T_h$. Thus $\rho = 0\ \forall\ h < s_a$ and $> e_a$, as its waiting latency lies in valid allowable interval. Therefore, waiting cost rises power usage when it is planned at later hours. So, waiting price directly proportional to ζ_a.

A resident can adjust the waiting cost by changing ζ_a on all appliances in the house. By taking $\zeta_a = 1$ for "a" $\in A_p$, the load regulator process tries to decrease the power charges, the d_a for every appliance "a" as

$$d_a = \frac{\varrho_a - s_a}{e_a - s_a} \tag{6.17}$$

With the increase of ζ, electricity bill increases with decrease in delay time is shown in Fig. 6.6, chosen limits for ζ are 1 and 1.01. Thus importance is being given to the residents as a result of which they will be able to optimize energy usage according to their needs.

The objective is to control energy usage, also to find optimal energy scheduling of appliances which reduces power bill and lower DF.

Using Eqs. (6.1) and (6.2), the electricity bill can be determined in terms of energy use, as shown in Eq. (6.4). Energy usage scheduling that is both cost-effective and efficient can be depicted as:

$$\text{Minimize}(w_1.P_{\text{sch}} + w_2.\text{DF}_a) \tag{6.18}$$

Subject to

$$t_a \in [s_a, e_a - \text{lt}_a] \tag{6.18a}$$

$$\text{ont}_a \in [\text{ssch}_a, \text{esch}_a - \text{ltsch}_a] \tag{6.18b}$$

$$P_{\text{sch}} = \sum_{h=1}^{24} p_{\text{tch}}\, p^h_{\text{sch}} \cdot p^h_{\text{sch}} - P^{\text{out}}_{\text{max}} \tag{6.18c}$$

$$\text{DF}_a = \frac{\zeta_a^{e_a - h}}{p_t^{T_h}} .x_a^h \tag{6.18d}$$

where, mathematical statement 18 gives the objective of reduction of costs, which is demonstrated as a binary programming problem. The parameters w_1, w_2 are used to control the load and delay cost given in Eq. (6.18). Eqs. (6.18a) and (6.18b) denote the horizon of scheduling for RTP and RTP + IBR, where It is possible to schedule the load. Eq. (6.18c) estimates energy consumption of appliance with anticipated scheduling approach according to its usage with RES integration for cost savings and user independence. Therefore, the scheduling optimization problem on energy consumption can be expressed in Eq. (6.18d), where $\text{DF}_a \in [0, 1]$.

6.4 Genetic algorithm

The primary objective of an optimization algorithms is choosing suitable energy models to fulfill the requirements such as exactness, reliability, convergence and efficiency. Generally, available optimized processes are divided into three categories namely, deterministic, stochastic, and fuzzy algorithms. A proposed method has chosen genetic algorithm (GA) to manage energy in an adjusted mode to accomplish prosumer's favorites.

6.4.1 Literature survey

In the field of SG, researchers have been trying to distribute load, to schedule appliances, to reduce total cost, maximize user comfort, and ensure optimal energy sharing among appliances. The objective of our chapter is to study different algorithms, aiming in decrease of electricity bill by scheduling of appliances.

A dynamic programming method is being used in Sianaki, Hussain, and Tabesh (2010) proposed technique to reach the consumer favorite through for an effective, optimal, and preferred solution during peak hours. Appliances are scheduled based on user's appliances during peak to nonpeak hours. He categorized the appliances into four categories according to Emergency, Safety, Power Cost, and Entertainment. He used knapsack approach for scheduling and to achieve optimized decision making in smart appliances, used AHP to rank them, can allow customers to control their appliances to save money on their electricity bill or to attain a specific level of comfort during peak hours. The author has done the experimentation on very short data collected from limited number of smart appliances in a smart meter.

Rathnayaka, Potdar, and Kuruppu (2011) have proposed a standard procedure in artificial data set preparation for indistinguishable profiles of prosumers for experimentation. The realistic energy behavior of the prosumer in various seasons like winter and summer is considered for synthetic data generation. It contains energy generation and consumption in time series for Australia. The data presents that prosumer energy-sharing characteristics are dynamic as it oscillates in accordance with time. The produced, consumed, and shared synthetic data set were generated for a month from 500 prosumers. It is used in the derivation of a new data set with determined frequency that could be lessened to hourly basis.

Kumaraguruparan, Sivaramakrishnan, and Sapatnekar (2012), used dynamic pricing paradigm which is intended to enable domestic energy management by scheduling appliances, with a focus on day-ahead pricing This pricing divides appliances into three categories: must run, fixed, and flexible scheduler, with the goal of managing energy more evenly during peak hours. These appliances are mapped into a multiple knapsack problems to reduce electricity bills and achieve energy usage homogeneity. LINDO API is used in method performance testing and, observed this technique is beneficial in introducing uniformity to a consumer's energy profile, but failed to satisfy the user comfort.

Miao, Huang, and Chen (2012) proposed an intelligent scheduling technique using GA, suitable for smart home to diminish the energy expenses through scheduling appliances based on starting time, estimated from power intakes and operation period constraints. In SG atmospheres, Advanced Metering Infrastructure could inevitably plan appliances operation time to lessen total power usage by fulfilling appliance operation constraint such as starting time should be lie in two predefined time instances, and energy system should not be burdened in any instance. The efficiency of the method is compared with SA and greedy search technique. Results describe that GA based scheduling algorithm perform proficiently and cuts electricity bill.

Sianaki and Masoum (2013) developed a multiattribute decision-making method for fine-tuning and regulating energy consumption and cost. However, this strategy ignores the prosumer's consumption habits.

Carpinelli et al. (2013) proposed tool intended to analyze potential benefits of distributed electricity storages in SGs with reference to both conceivable regulatory schemes and services offered. The SG operator assumed to have ownership on operation of the energy storage arrangements, where energy storage devices and an external interconnected grid are explored.

Zhao, Lee, Shin, and Song (2013) focused on arrangement of energy management system in a home area network based on SG and then offer an well-organized scheduling process for energy usage. The Dr information is obtained by a Home gateway (HG), which reflects the present energy cost that is shifted to energy management controller (EMC). With Dr, EMC achieves best energy scheduling arrangement that can be transported to each appliance through HG. RTP model is adopted, work on more appliances, all through the period with minimum electricity cost, but, it may harm energy system due to high PAR. But author had used RTP + IBR pricing model that would successfully diminish both energy cost and PAR, thereby, improved the stability of energy system.

Oladeji and Olakanmi (2014) has proposed heuristic optimization method which rearranges the appliances with certain purposes such as intensification of load factor, drop PAR or energy cost. The objective is to lessen overall energy cost by adopting a dynamic pricing despite the fact to satisfying a desired constraint. The conclusions prove that GA can minimize energy utilization over defined constraints, thus reduce total electricity bill in a smart pricing environment for the Nigerian consumer.

Ten, Yessenbayev, Shamshimova, and Khakimova (2015) has integrated Smart house's with Renewable energy sources like electricity heating subsystems and solar thermal heater respectively. It uses either electrical or thermal energy based on the state of accumulator battery stack and temperature present in heating system. The objective of this arrangement uphold the system state with in permitted range and also to lessen the expenditures for power utilization from utility grid, achieved by using genetic algorithm.

Razzaq, Zafar, Khan, Butt, and Mahmood (2016) proposed prosumer's energy generation and sharing technique using linear programming equations, single and multiple criteria. It permits users to sell the excessive energy with them to users with an energy deficiency in an optimized manner. This follows a simple process and use intelligent techniques to manage prosumer's energy sharing in a smart community.

Rasheed, Javaid, Ahmad, et al. (2016), binary knapsack problem is used in his methodology for appliances are scheduled using a dynamic pricing scheme. Higher priority is given for user comfort, and cost reduction is done on user preference. Author divided the appliances into user dependent, interactive schedulable, and unschedulable. Experimentation shows the result of lowering costs of approximately 17%.

Rasheed, Javaid, Awais, et al. (2016) proposed a technique using GA to diminish cost, PAR and to optimize energy usage based on user preferences. He categorized the appliances into five groups for optimization of thermostatically controlled appliances by preferring user comfort. Effectiveness is checked by the comparison of scheduled and unscheduled appliances. But some appliances are designed solely to save money, minimize PAR, and improve user comfort.

Javaid et al. (2017) proposed a HEM manager algorithm using four heuristic optimization procedures, namely, Bacterial Foraging, GA, Binary Particle Swarm optimization (BPSO) and Wind Driven Optimization. Furthermore, a combined method is proposed named as Genetic BPSO (GBPSO). Four techniques were tested with the attention of important household appliances in RTP. Outcomes illustrate that each algorithm lessens the energy cost and restricts the PAR. GA based HEM organizer accomplishes fairly best in terms of lower in PAR. Additionally, GBPSO method achieves efficiency than other algorithms in terms of lowering of electricity cost.

Rehman, Fatima, and Khan (2019) proposed a technique for achieving prosumer's various aims in a more efficient and intelligent manner. A GA is being utilized to share energy in the most efficient way possible without upsetting the preferences of the prosumers. The results show that it considerably reduces energy sharing losses while maintaining user preferences. The complete literature review is summarized in Table 6.2.

6.4.2 Proposed energy optimization technique

As we know GA is extensively accepted, GA optimization phases, includes, below are diagrams of the initiation, fitness function, selection, crossover probability (P_C) and mutation probability (P_M). The RES is to be used when demand of energy does not exceed the threshold limit, otherwise the GA is being used to find scheduling patterns when the RES failed to fulfill the energy demand. If the optimal results cannot be achieved with the present population, to find the fitness function in step-15. Steps 24–28 are terms that define the process of creating a new population from the chromosomes of parents. Using P_C and P_M, generate new population in an iteration because, convergence rate of the algorithm depends on both parameters. Complete description of the proposed algorithm is shown below.

In this chapter, the traditional average crossover is replaced by Inversed Bi-segmented Average Crossover (IBAX). In this approach, the chromosomes were divided X and Y are divided into two segments, and Z is

TABLE 6.2 Summary of literature works.

	Technique	Objective	Features	Limitations
Sianaki et al. (2010)	Knapsack	Determine when an increase is in electricity price rate.	Appliances priority criteria and alternatives are presented	Suitable to limited data of smart appliances
Rathnayaka et al. (2011)	Prosumer Community Group	Analyzation of prosumer participation and highlighted the open research issues.	A novel concept of goal-oriented virtual prosumer communities	Only gave concept of prosumer community group
Kumaraguruparan et al. (2012)	Multiple Knapsack Method	Minimize customer's energy cost by scheduling appliances under day-ahead variable peak pricing.	Schedule residential appliances energy under a dynamic pricing model.	User comfort was ignored
Miao et al. (2012)	Genetic Evolutionary Task Scheduling	Reduce energy price	Novel GA optimization technique is proposed	Only cost function is formulated
Sianaki and Masoum (2013)	Fuzzy TOPSIS	Maximize their participation in demand response programs	Fuzzy rules for analyzing and anticipate users' consumption.	Real data was not used and not integrated renewable energy.
Carpinelli et al. (2013)	Distributed power preserving devices	Minimize reactive power and energy cost	To reduce cost and reactive power, four distinct integration scenarios of DESSs and capacitors are offered.	PAR and user comfort was ignored.

Zhao et al. (2013)	Distributed Energy Storage	Minimize PAR and cost.	The problem of optimization is stated, and the time slot split is altered to regulate cost and PAR.	Increase in System complexity.
Oladeji and Olakanmi (2014)	Genetic Algorithm	Reduce energy price	Mathematically specified objective function and limitations	Considered only the cost
Ten et al. (2015)	Genetic Algorithm	Minimize energy usage and price	Mathematical modeling of electrical, thermal and hybrid system	Suitable for single house.
Razzaq et al. (2016)	Linear Programming	Proposed Prosumer Energy Sharing and Management technique for Demand Side Management	Users can trade their unused energy with other users who are experiencing a power outage.	There is no automated method for determining user priority.
Rasheed, Javaid, Ahmad, et al. (2016)	Constraint solving techniques	Optimal scheduling method to lessen energy price	Optimization is achieved using TOU and RTP pricing.	Prosumer concept is ignored
Rasheed, Javaid, Awais, et al. (2016)	Genetic Algorithm	Determine household appliances which consumed the energy optimally to reduce electricity bill.	Increase savings on electricity bill and user comfort.	Prosumers are ignored and GA applied to HVAC only
Javaid et al. (2017)	Heuristic optimization	A hybrid algorithm which is Genetic BPSO (GBPSO)	In terms of cost and PAR restriction, the HEM controller outperforms existing algorithms-based HEM controllers.	Prosumers not considered
Rehman et al. (2019)	Genetic Algorithm	Optimize prosumers energy sharing in	Significantly improves the loss of energy sharing.	

created from X and Y by averaging genes from each section. whereas, in IBAX segment X and Y by separating total genes into two and make sure that both the segments must have an equal number of genes in an even count. select first segment, derive gene in Z by inversely pairing X and Y. Repeat this pairing till last gene of X and first gene of Y have coupled, compute average on paired genes to calculate Z. Detailed IBAX algorithm is shown below.

Algorithm: Pseudocode for proposed Algorithm

1. begin
2. set the probability of mutation $\partial \leftarrow 0.005$
3. randomly initialize all patterns of appliances P_0 of dimension $|P|$
4. choose top chromosome from $P_0 \cup \overline{P_0}$, where $\overline{P_0}$ is the complement of P_0
5. set $q=1$ and $\delta=$ False
6. while(($q<$ epoch) && ($\delta \neq True$))
7. j = 0
8. for time = 1 to 24 do
9.
10. if ($P_a \le$ RES) then
11. appliance "a" turn ON
12. else if ($P_a >$ RES) then
13. for i = 1 to $|P|$ do

$$\mathrm{Ft} = w_1.P_{sch} + w_2.\mathrm{DF}_a$$

14. if ($\mathrm{Ft}_i \le \mathrm{Ft}_{i-1}$) then
15. satisfy all constraints Equations

$$t_a \in [s_a, e_a - \mathrm{lt}_a]$$

$$\mathrm{ont}_a \in [\mathrm{ssch}_a, \mathrm{esch}_a - \mathrm{ltsch}_a]$$

$$P_{sch} = \sum_{h=1}^{24} p_{tch}\, p_{sch}^{h} \cdot p_{sch}^{h} - P_{max}^{out}$$

$$\mathrm{DF}_a = \frac{\varsigma_a^{e_a - h}}{P_t^{T_h}} x_a^h$$

16. check appliances status
17. turn on appliance "a"
18. else
19. go to step 15
20. select two pairs of chromosomes in random
21. perform crossover and mutation
22. generate P_{new}
23. go to step 15
24. return optimal solution

Algorithm: IBAX Algorithm

Step 1: choose two parents called as X and Y.

Step 2: check whether the dataset is odd or even by counting genes in chromosome.

Step 3: go to step 9 when count is odd

Step 4: bipartite total genes of two chromosomes (X and Y) separately, make sure that X and Y have equally shared genes.

Step 5: extract first segment, calculate Z(offspring) for each gene pair first gene from X to last gene in Y.

Step 6: repeat step 5 until pairing last gene in X and first gene in Y have inversely mated and then calculate Z by averaging their values.

Step 7: extract the second segment, calculate Z for each gene by inversely pairing first gene from X with gene on Y.

Step 8: repeat step 7 till pairing last gene and first gene of X,Y and have produced Z by calculating average of pair values. Goto step 10

Step 9: in odd datasets, last genes of chromosomes will not paired with second segment, will automatically be mated with each other to produce offspring.

Step 10: return offspring Z

The rate of convergence is very important when we are working with a sequence of approximation. In general, the proposed method's convergence is assessed in accordance with chosen parameters and decided constraints. In general, the first operation in GA is to take a random sample of the population. The primary goal is then assessed using the criteria provided. In general, many researchers have chosen roulette wheel selection criteria, but the proposed technique have used IBAX selection cross over operation, single using P_C and P_M. With the increase in value of P_C, there is an increase in convergence rate, and high worth of P_M It's possible that this will lower the quality of the optimal solution and, as a result, produce premature convergence. Thus, we tested the impact by setting different values to P_C and P_M and chosen a best value which ensures fast convergence and optimality. Furthermore, in order to avoid premature convergence (Askarzadeh, 2018; Wang, Li, Maharjan, Zhang, & Guo, 2018), a sigma scaling factor is being used, it tries to moderate the selection pressure over time and in accordance with fitness scores (Time has Arrived for Time Variant Pricing, 2018). we conduct simulations, by choosing residence = 1 and Ap = 6. From the graph we can observe that for 200 iterations, the P_{tch} value of algorithm has chosen RTP + IBR based energy pricing on load scheduling. Up to 300 repetitions, $P^{\text{out}}_{\text{max}}$ value has converged for load balancing using RES. Meanwhile, due to short time intervals and the addition of manually operated appliances, algorithm convergence is lacking, despite the fact that it appears to be reasonably high.

6.5 Results and discussion

Table 6.3 shows control parameters chosen in GA algorithm. Experimentation results are described based on Dr. programs that governs residential load. In general, the experimentation was performed on unscheduled and scheduled cases. The class of users do not have any smart technology-based devices like smartphones, laptops and tabs and no knowledge of load scheduling is called as conventional users. This class of users are considered as unscheduled class as they don't have any energy management/planning/scheduling, no price awareness. Whereas the other class of users called as smart users, who have smart home appliances which can operate on smart technology. Therefore, smart users have HEM and RTP data in their homes through home display. The experimentation results are attained intended for four scenarios (1) Unscheduled, (2) setting up with RTP, (3) setting up with RTP + IBR and (4) Scheduled with RES. In order to verify chosen load management effectiveness and energy pricing are considered in work, results after completion of experimentation shows all four scenarios are studied and reflected.

1. Unscheduled: This conventional user usually pays more electricity bill because of no knowledge about the load management in accordance with hourly varying electricity price. The maximum cost in every hour is $\text{cost}_{\text{Max}}^{\text{Unsch}} = \$337.7$.
2. Scheduled with RTP: In this category, we use GA mechanism for scheduling with RTP in order to manage their electrical appliance loads.

TABLE 6.3 Parameters of genetic algorithm.

Parameter	Value
Population type	Fixed size
Population size	10 individuals
Number of generations	800
Selection	IBAX rule
Crossover rate	1/generation
Crossover points	1
Mutation rate	0.5/generation
Mutation points	1
Objective	See relation
Ending criteria	After 800 generation

3. Here, we considered a GA based scheduling RTP for a single day, with the goal of limiting bill payment by restricting energy usage. Hence, the hourly cost can be calculated as:

$$\text{cost}_{\text{tot}}^{T} = \sum_{a_i \in A_{p,\text{sch-Load}}} \text{Load}_{a_i,\text{sch}} \times \text{EP} \tag{6.19}$$

Our objective is to minimize cost in each hour, so, it is described as min $(\text{cost}_{\text{tot}}^{T})$

Further, the limit values considered for $C_1 : \text{cost}_T^{\text{sch}}, C_2 : \sigma_{\text{load}}$ and $C_3 : \text{load}_T^{\text{sch}}$ are [0.2, 337.7], $\leq$ 1859.2 and (Ayan & Turkay, 2018; Khalid et al., 2018).

Scheduled with RTP + IBR: It uses GA method, with a combined pricing model for optimized load administration. The objective remains same as pricing model with RTP, however, because of the tariff pricing, the restrictions for improving the reliability of a power system have altered. So, the cost per hour is defined as:

$$\text{cost}_{\text{tot}}^{T} = \sum_{a_i \in A_{p,\text{sch-Load}}} \text{Load}_{a_i,\text{sch}} \times \text{EP} \tag{6.20}$$

Our motive is to minimize electricity consumption cost in each hour, described as min $(\text{cost}_{\text{tot}}^{T})$

Further, the limit values considered for $C_1 : \text{cost}_{\text{tot}}^{T}, C_2 : \sigma_{\text{load}}$ and $C_3 : \text{load}_T^{\text{sch}}$ are [0.2,279], $\leq$ 1474.7 and [1.5,9.5].

Table 6.4 shows the costs based on a variety of load and price profiles of RTP and RTP + IBR.

4. Scheduled with RES: In this class, the users will get the benefits of different pricing plan and also the integration of RES framework lessens the load in turn limit the cost. This framework reduces power cost remarkable as the RES integration move the peak load onto the RES.

TABLE 6.4 Cost estimates based on various load and price profiles RTP and RTP + IBR.

	RTP			RTP + IBR		
Event	Load (kWh)	EP ($/kWh)	Cost ($)	Load (kWh)	EP ($/kWh)	Cost ($)
Min − Load, Min − EP	1	10.2	10.2	1.5	10.2	15.3
Min − Load, Max − EP	1	30.7	30.7	1.5	30.7	46.5
Max − Load, Min − EP	11	10.2	112.2	9.5	10.2	112.2
Max − Load, Max − EP	11	30.7	337.7	9.5	30.7	290.11

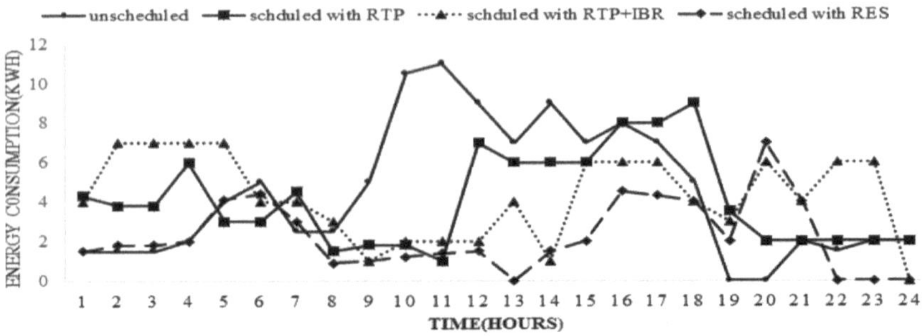

(A)

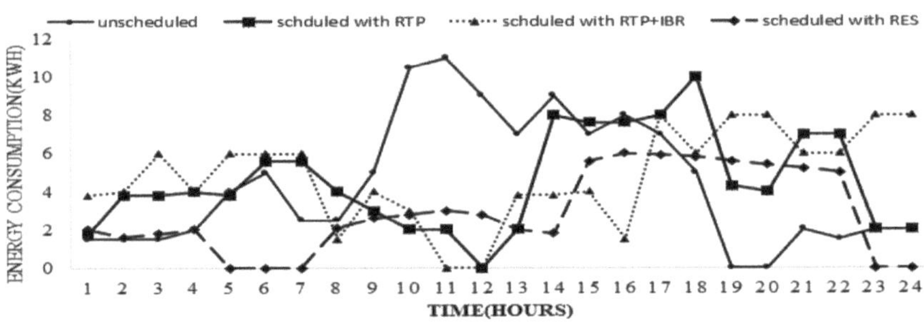

(B)

FIGURE 6.7 (A) Unscheduled and scheduled load profiles for a variety of scenarios; RTP, RTP + IBR, and RES without a control parameter. (B) Unscheduled and scheduled load profiles for a variety of scenarios, including RTP, RTP + IBR, and RES with a control parameter.

In unscheduled load consumption process, without using any load management techniques, more bill is imposed to the conventional users more bills as a result of unscheduled load use, without taking market prices into account. Fig. 6.7A depicts load patterns that are both unplanned and scheduled but are not controlled by a load control parameter. Unscheduled peak load is moved to off-peak hours, according to our findings. Fig. 6.7B, on the other hand, depicts the unplanned and scheduled load patterns, as well as the control parameter. In comparison to traditional users, this type of user is more worried about electricity costs. They are well aware of the importance of good connectivity with DSM programs and the numerous scheduling options available to retailers. From the figures we can observe the load and the electricity prices have shown the maximum values as there is not constraint in energy consumption.

Unscheduled peak load is moved to off-peak hours, as seen in Fig. 6.8A. We can see from the graph that there is an unscheduled peak load and an extraordinary peak cost since traditional users are willing to pay more fees.

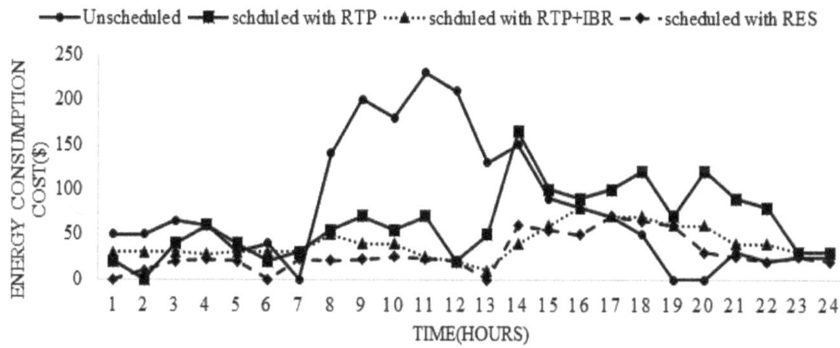

(A)

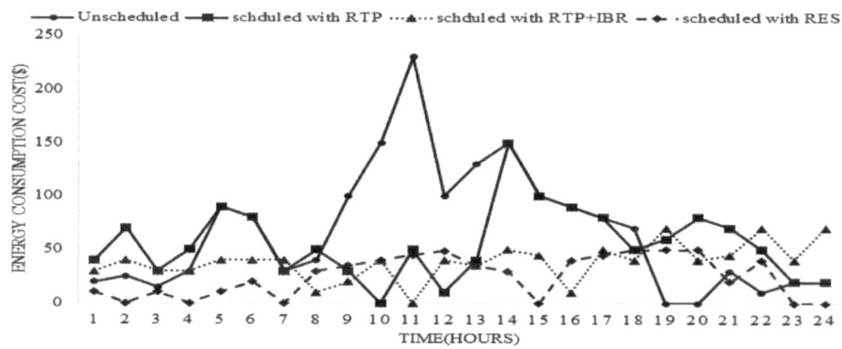

(B)

FIGURE 6.8 (A) Cost profiles for unscheduled and planned scenarios; RTP, RTP + IBR, and RES without control parameter. (B) Cost profiles for unscheduled and planned scenarios; RTP, RTP + IBR, and RES with control parameter.

However, the PAR value in the unscheduling method is higher than in other scheduling schemes. The maximum cost has been reduced from 225 to165$/kWh. Furthermore, although employing RTP results in a total electricity cost of $165, using RTP + IBR results in a cost of around $88. The electricity cost is further reduced as a result of the control parameter's influence, as shown in Fig. 6.8B, with defined user comfort. However, the stability of the systems may be jeopardized due to rebound peaks caused by load shifting. As a result, load management lists that rely solely on RTP appear less practical in situations where grid dependability and user satisfaction are also necessary. For this, we'll need a mechanism that reduces both energy costs and user annoyance while maintaining high system stability. With the shortest possible delay time, the electricity It is possible to achieve both cost and user comfort. The RTP + IBR approach was discovered to save up to $77/day. Further, when we incorporate renewable energy sources (RES) into our model, we can reduce costs and customer discomfort while also boosting system stability. Because there is another source of energy, consumers do not have to rely solely on grid electricity and can combine the two for cost savings. We can see from both statistics that the overall cost of RTP + IBR is $88, which is further lowered to $51 with the effect of and integration of RES.

Overall unscheduled cost and the PAR values were shown in Fig. 6.9A and B. From the figures, it can be understood that in all scheduled scenarios, the cost of energy use with is reasonably low. Furthermore, the cost of employing RES and IBR + RTP is significantly lowered. Fig. 6.9B shows that adopting the proposed approach and effect of PAR was decreased from 2.68 to 1.5. It shows that the PAR value has decreased from 1.23 to 1.12, and that by including RES, we were able to achieve optimum user comfort. Also demonstrates that by utilizing RES, we can eliminate user reliance on the grid for cost-effective benefits as well as a reduction in PAR while achieving the most thrilling user comfort and the shortest delay time.

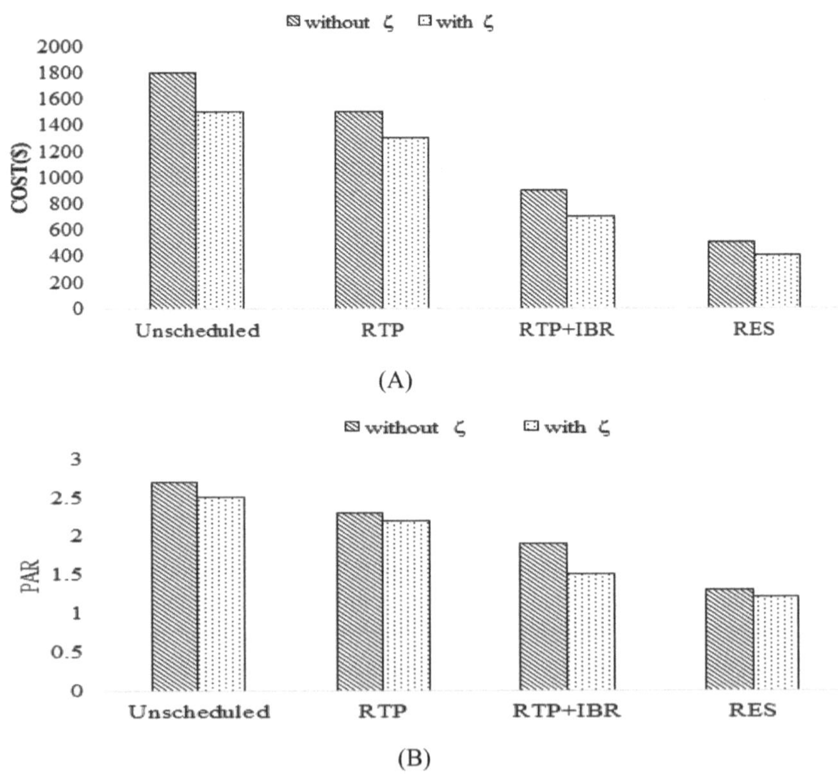

FIGURE 6.9 (A) Energy cost estimation without and with control parameter. (B) PAR estimation without and with control parameter.

6.6 Conclusions

A house energy optimization approach is proposed with addition of RESs and ζ. The energy consumption and waiting cost was balanced, by regulating the, consumers have the option of prioritizing their comfort over the price of electricity. The performance was evaluated by varying the resource load and categorizing the results according to the user requirements. Further, mathematical models with respect to loads have been formulated, primarily, customer energy consumption cost was minimized using RTP. However, it has significantly reduced PAR. Further, the method was revised adopted RTP + IBR, a joint pricing model for electricity price used to reduce cost and PAR. But user comfort is omitted ultimately. Later, we integrated RES integrated with supplementary Electricity price is considered a critical limitation for scheduling load based on user priority. Finally, GA is utilized to measure performance by directing all test cases. The suggested algorithm has the ability to manage residential load while maintaining user comfort, as demonstrated by simulation results. The cost of energy pricing has been cut from \$225 to \$51 and the PAR has been reduced from 2.68 to 1.12.

References

Abushnaf, J., & Rassau, A. (2018). Impact of energy management system on the sizing of a grid-connected PV/Battery system. *Electronic Journal, 31*, 58–66.

Alam, M. R., St-Hilaire, M., & Kunz, T. (2016). Computational methods for residential energy cost optimization in smart grids: A survey. *ACM Computing Surveys (CSUR), 49*(1), 2.

Asgher, U., Rasheed, M., & Awais, M. (2018). Demand response benefits for load management through heuristic algorithm in smart grid. *International Symposium on Recent Advances in Electrical Engineering (RAEE)*, 1–6. Available from https://doi.org/10.1109/RAEE.2018.8706886.

Askarzadeh, A. (2018). A memory-based genetic algorithm for optimization of power generation in a microgrid. *IEEE Transactions on Sustainable Energy, 9*, 1081–1089.

Ayan O., & Turkay, B. (2018). Domestic electrical load management in smart grids and classification of residential loads. In *2018 5th International conference on electrical and electronic engineering (ICEEE)* (pp. 279–283). doi:10.1109/ICEEE2.2018.8391346.

Barbato, A., & Capone, A. (2014). Optimization models and methods for demand-side management of residential users: a survey. *Energies, 7*(9), 5787–5824.

Baronti, F., Vazquez, S., & Chow, M. (2018). Modeling, control, and integration of energy storage systems in E-transportation and smart grid. *IEEE Transactions on Industrial Electronics, 65*, 6548–6551.

Bui, V., Hussain, A., & Kim, H. (2018). A multiagent-based hierarchical energy management strategy for multi-microgrids considering adjustable power and demand response. *IEEE Transactions on Smart Grid, 9*, 1323–1333.

Carpinelli, G., Celli, G., Mocci, S., Mottola, F., Pilo, F., & Proto, D. (2013). Optimal integration of distributed energy storage devices in smart grids. *IEEE Transactions on Smart Grid, 4*(2), 985–995.

Dahlke, S. (2020). Integrating energy markets: Implications of increasing electricity trade on prices and emissions in the western United States. *International Journal of Sustainable*

Energy Planning and Management, *25*, 45–60. Available from https://doi.org/10.5278/ijsepm.3416.

Erdinc, O., Tascikaraoglu, A., Paterakis, N. G., Eren, Y., & Catalao, J. (2017, July 16–20). End-user comfort oriented day-ahead planning for responsive residential HVAC demand aggregation considering weather forecasts. In *Proceedings of the 2017 IEEE power & energy society general meeting*. Chicago, IL.

Good, N. (2018). A transactive energy modelling and assessment framework for demand response business cases in smart distributed multi-energy systems. *Energy*.

Hongbo, Z., Yan, G., & Yong, H. (2018). Real-time pricing for demand response in smart grid based on alternating direction method of multipliers. *Mathematical Problems in Engineering*, *10*, 2018. Available from https://doi.org/10.1155/2018/8760575, Article ID 8760575.

Javaid, N., Naseem, M., Rasheed, M., Mahmood, D., Khan, S. A., Alrajeh, N., & Iqbal, Z. (2017). A new heuristically optimized home energy management controller for smart grid. *Sustainable Cities and Society*, *34*, 211–227.

Javaid, N., Naseem, M., Rasheed, M. B., Mahmood, D., Khan, S. A., Alrajeh, N., & Iqbal, Z. (2017). A new heuristically optimized home energy management controller for smart grid. *Sustainable Cities and Society*, *34*, 211–227.

Jindal, A., Singh, M., & Kumar, N. (2018). Consumption-aware data analytical demand response scheme for peak load reduction in smart grid. *IEEE Transactions on Industrial Electronics*, *65*, 8993–9004.

Khalid, A., Javaid, N., Guizani, M., Alhussein, M., Aurangzeb, K., & Ilahi, M. (2018). Towards dynamic coordination among home appliances using multi-objective energy optimization for demand side management in smart buildings. *IEEE Access*, *6*, 19509–19529.

Khan, A. R., Mahmood, A., Safdar, A., Khan, Z. A., & Khan, N. A. (2016). Load forecasting, dynamic pricing and dsm in smart grid: A review. *Renewable & Sustainable Energy Reviews*, *54*, 1311–1322.

Kumaraguruparan, N., Sivaramakrishnan, H., & Sapatnekar, S. S. (2012). *Residential task scheduling under dynamic pricing using the multiple knapsack method. 2012 IEEE PES innovative smart grid technologies (ISGT)* (pp. 1–6). IEEE.

Luo, C., Huang, Y., & Gupta, V. (2018). Stochastic dynamic pricing for EV charging stations with renewable integration and energy storage. *IEEE Transactions on Smart Grid*, *9*, 1494–1505.

Mary, G. (2014). Smart grid cost optimization using genetic algorithm. *International Journal of Research in Engineering and Technology*, *3*, 282–287. Available from https://doi.org/10.15623/ijret.2014.0319051.

Miao, H., Huang, X., & Chen, G. (2012). A genetic evolutionary task scheduling method for energy efficiency in smart homes. *International Journal of Electrical Engineering*, *7*(5), 1827–1835.

Oladeji, O., & Olakanmi, O. (2014). A genetic algorithm approach to energy consumption scheduling under demand response. In *IEEE 6th international conference on adaptive science and technology (ICAST)* (pp. 1–6).

Rasheed, M., Javaid, N., Ahmad, A., et al. (2016). Energy optimization in smart homes using customer preference and dynamic pricing. *Energies*, *9*(8), 593.

Rasheed, M., Javaid, N., Awais, M., et al. (2016). Real time information-based energy management using customer preferences and dynamic pricing in smart homes. *Energies*, *9*(7), 542.

Rasheed, M. B., Javaid, N., Ahmad, A., Awais, M., Khan, Z. A., Qasim, U., & Alrajeh, N. (2016). Priority and delayconstrained demand side management in real-time price environment with renewable energy source. *International Journal of Energy Research, 40*, 2002–2021.

Rathnayaka, A. D., Potdar, V. M., Dillon, T., & Kuruppu, S. (2015). Framework to manage multiple goals in community-based energy sharing network in smart grid. *International Journal of Electrical Power & Energy Systems, 73*, 615–624.

Rathnayaka, A. D., Potdar, V. M., & Kuruppu, S. J. (2011). *An innovative approach to manage prosumers in smart grid. 2011 World congress on sustainable technologies (WCST* (pp. 141–146). IEEE.

Razzaq, S., Zafar, R., Khan, N., Butt, A., & Mahmood, A. (2016). A novel prosumer-based energy sharing and management (pesm) approach for cooperative demand side management (DSM) in smart grid. *Applied Sciences, 6*(10), 275.

Rehman, Z., Fatima, M., Khan, A., et al. (2019). Energy optimization in a smart community grid system using genetic algorithm. *International Journal of Communication Systems*, e4265. Available from https://doi.org/10.1002/dac.4265.

Sanjeev, P., Padhy, N. P., & Agarwal, P. (2018). Peak energy management using renewable integrated DC microgrid. *IEEE Transactions on Smart Grid, 9*, 4906–4917.

Sarkar, S., Chakrabarti, U., Bhattacharyya, S., et al. (2020). A comprehensive assessment of the need and availability of smart grid technologies in an electricity distribution grid network. *Journal of the Institution of Engineers (India): Series B, 101*, 753–761. Available from https://doi.org/10.1007/s40031-020-00486-1.

Sianaki, O. A., Hussain, O., & Tabesh, A. R. (2010). *A knapsack problem approach for achieving efficient energy consumption in smart grid for end users' life style. 2010 IEEE conference on innovative technologies for an efficient and reliable electricity supply* (pp. 159–164). IEEE.

Sianaki, O. A., & Masoum, M. A. (2013). *A fuzzy TOPSIS approach for home energy management in smart grid with considering householders' preferences. 2013 IEEE PES innovative smart grid technologies conference (ISGT)* (pp. 1–6). IEEE.

Tan, K. M., Ramachandaramurthy, V. K., & Yong, J. Y. (2016). Integration of electric vehicles in smart grid: A review on vehicle to grid technologies and optimization techniques. *Renewable & Sustainable Energy Reviews, 53*, 720–732.

Ten, V., Yessenbayev, Z., Shamshimova, A., & Khakimova, A. (2015). Optimized smallscaled hybrid energy management of a smart house based on genetic algorithm. In *IEEE 14th international conference on machine learning and applications (ICMLA)*, (pp. 1153–1158).

Time has Arrived for Time Variant Pricing but What Kind? (2018). Available online: http://www.menloenergy.com/?p = 349 (accessed 23.08.18).

Wang, K., Li, H., Maharjan, S., Zhang, Y., & Guo, S. (2018). Green energy scheduling for demand side management in the smart grid. *IEEE Transactions on Green Communications and Networking, 2*, 596–611.

Yaagoubi, N., & Mouftah, H. T. (2015). User-aware game theoretic approach for demand management. *IEEE Transactions on Smart Grid, 6*, 716–725.

Zhao, Z., Lee, W. C., Shin, Y., & Song, K. (2013). An optimal power scheduling method for demand response in home energy management system, in*IEEE Transactions on Smart Grid, 4*(3), 1391–1400. Available from https://doi.org/10.1109/TSG.2013.2251018.

Zhao, Z., Lee, W. C., Shin, Y., & Song, K. (2013). An optimal power scheduling method for demand response in home energy management system. *IEEE Transactions on Smart Grid, 4* (3), 1391–1400.

Zhao, Z., Lee, W. C., Shin, Y., & Song, K. (2013). An optimal power scheduling method for demand response in home energy management system. *IEEE Transactions on Smart Grid, 4*, 1391–1400.

Chapter 7

Distributed semantic architecture for smart grids: an industrial approach

D. Siva Krishna[1], B. Raja Koti[2] and P. Muralidhara Rao[3]
[1]*Department of CSE, GMR Institute of Technology, Srikakulam, India,* [2]*Department of CSE, GITAM Institute of Technology, Vishakhapatnam, India,* [3]*School of Computer Science and Engineering, Vellore Institute of Technology, Vellore, India*

7.1 Introduction

7.1.1 Introduction to Industry 4.0

Companies are spending heavily on technologies to improve user-friendliness by using artificial machines that solve many real-world problems. There are machines to be integrated as a single component. In this connection, technology growth improves the age of industrial growth. Industry 4.0 is a paradigm to link many types of physical equipment to the internet via electronics. It is critical to adopt both vertical and horizontal strategies. Ten digital industrial technologies refer to the convergence and application of Industrial 4.0.

They are:

1. robotics applications
2. data Science
3. computer vision
4. vertical and horizontal integration
5. cloud computing
6. cyber security
7. big data analytics
8. industrial internet of things
9. blockchain technology
10. quantum computing

In Industry 1.0, the data is generated through web pages called static data. In the next version of 2.0, data flows in a dynamic nature called

Sustainable Networks in Smart Grid. DOI: https://doi.org/10.1016/B978-0-323-85626-3.00010-7

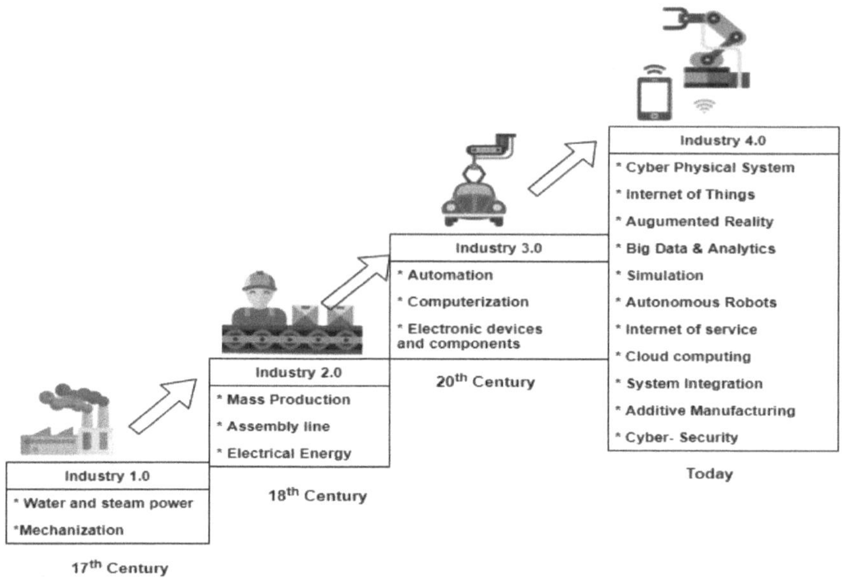

FIGURE 7.1 Improvements from Industry 1.0 to 4.0.

"Dynamic pages." In a later version, 3.0, data is analyzed using various tools to generate user friendly interactions between web pages. Now, the major goal of Industry 4.0 is to increase the efficiency and efficacy of automation operations. The 4.0 concept is an industrial paradigm that encompasses a series of future industrial advances, including cyber-systems, blockchain systems, and big data analytics. It is essential to develop and adopt all of these technologies for more intelligent manufacturing processes. The following Fig. 7.1 shows the improvements from traditional industry 1.0 to latest version of industry 4.0.

Industry 4.0 is self-organized for organizations that effectively utilize dynamic control in various aspects. There are forms of cycles to aggregates in the global manufacturing of lifelong goods. In future industrials, production characteristics will provide a level of organizing the things of innovative services. Hence, this will produce autonomous management, affiance degradation, consumption of resources, and service optimization, which leads to reduction and optimization of costs (Jadhav, Lokhande, & Gohokar, 2016).

7.1.2 Introduction to distributed systems

A distributed system is a system of heterogeneous computing elements that connect a network channel and spread data using moderators that perform operations autonomously but appear as a single coherent and integrated system to users (Yun, Junjie, Ji, & Hua, 2017). The computing element is generally referred to as a node. It can be either a software process or a hardware device.

Distributed systems (DSs) generally consist of two components:

- hardware concepts
  - multiprocessors
  - multicomputer
- software concepts
  - distributed real-operating system
  - network operational system
  - middleware technology tools (MTT)

Distributed systems have the following characteristics:

1. A number of self-contained nodes/components from which users cannot share information.
2. Access to limited resources, and distributed systems with several points of control and failure.
3. In parallel computing, software runs in parallel processes on multiple processors.

Characteristics of the system that are used to assess DS:

1. Sharing resources describes how users provide and use resources, as well as how they interact with one another.
2. Openness is defined as the facility of distributed system to be extended and improved.
3. The ability to uses malware that uses both hardware as well as software anywhere in the system can be used to assess the distributed systems.
4. *Transparency*: It is an abstraction from the application programmer and the clients that access system components in a single unit rather than the process of individual independent nodes that have been integrated in the distributed environment. It has different dimensions that are identified by ANSA.
   a. *Access transparency*: Allows remote information components to be accessed using identical operations.
      For example, file system operations in NFS.
   b. *Location transparency*: Allows information components to be accessed without knowledge of their location agents, for example, Pages in the Web.
   c. *Concurrency transparency*: Concurrently facilitates the operation of several processes without interference between them using shared information of objects, for example, database management system.
   d. *Failure transparency*: Abstraction of faults and allowing applications and users to complete their tasks besides the failure of individual nodes.
      For example, database distributed system

5. *Heterogeneity*: In a heterogeneous environment, the nodes in the network are accessed through services of a distributed nature.
6. *Middleware*: It provides the facility that functionality of the application is accessed through middleware technologies underlying with networks and its heterogeneities, namely hardware, software, and operating systems.

   For example, Common Object Request Broker Architecture (CORBA)
7. *Security*: Security plays a primary and important role in any system. Providing major elements as confidentiality, integrity, and availability.

   Confidentiality provides unauthorized individuals, integrity protects information from unauthorized accesses, whereas availability access the resources that are currently available.
8. *Failure handling/fault tolerance*: Sometimes systems fail and when a failure occurs in software or hardware components of a system, the results may be incorrect or the system may stop abruptly (Santos, Pinto, Vale, & Corchado, 2019). Failures in parallel distributed systems that is, some agents continue though there is a failure. Fault tolerance is achieved by maintaining the recovery and redundancy of nodes. Handling failure involves detecting, masking, and tolerating failures. Recovery from failures: involves software design such that saved points exist at intervals for recovery after a crash
9. *Concurrency*: In system/servers, multiple client requests can be processed in parallel for concurrent execution.
10. *Quality of service*: Provides required computing and communication resources to complete each task on time.

7.1.3 Introduction to distributed architectures

The distributed system architecture is a network that connects all components to transfer information and communicate with each other. The messages between the systems can be in the format of databases, objects, and files. The network must be connected using cables, IP addresses, or through a circuit board.

There are different types of distributed system architecture:

1. client-server architecture
2. three-tier architecture
3. multitier architecture
4. peer-to-peer architecture

Client-server architecture: This architecture consists of a server that shares resources like a database, printer, or server. It consists of multiple clients behind the computer. The mapped processors will not be necessarily 1:1. The Fig. 7.2 shows, a client should have an idea about the servers but servers need not have an idea about clients.

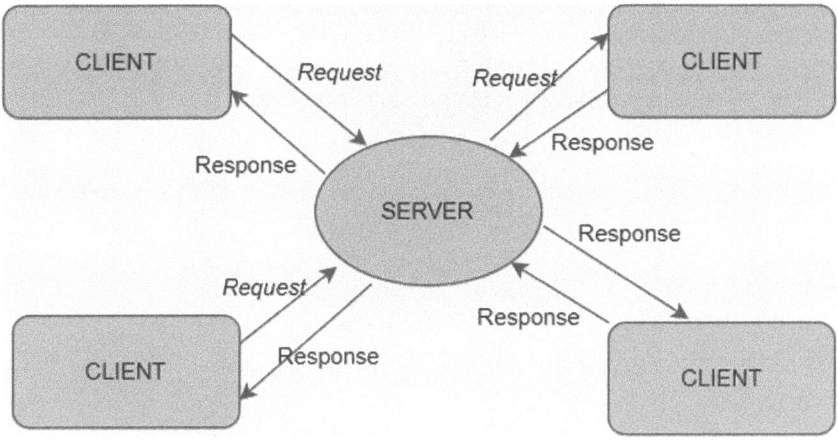

FIGURE 7.2 Client-server architecture.

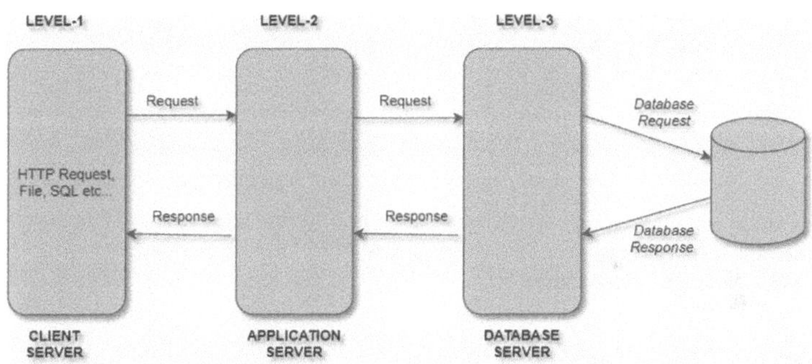

FIGURE 7.3 Three-tier architecture.

Three-tier architecture: In this architecture, each application layer will execute on a separate processor. Clients can rely on a middle tier for any decision making and processing need not be intelligent. Almost all first web applications fall under this category. The Fig. 7.3 shows, it is simpler to use a fat-client approach, which performs better than the thin-client systems. As demand increases, extra servers are added to this architecture to become scalable.

Multitier architecture: This architecture is an enterprise web service first created in n-tier and multitier architectures. In Fig. 7.4 shows, each server (tier 2 and 3) performs specialized services or tasks. This application server contains the business logic and interacts with both presentation tiers and data tiers.

Peer-to-peer architecture: In this architecture, we need not have any special machine or centralization to do the heavy lifting and intelligent work. These nodes contain equal participants to share data to individual nodes. Load is distributed among the machines responsible in making decisions in each client-server role. The Fig. 7.5 shows, the block chain architecture,

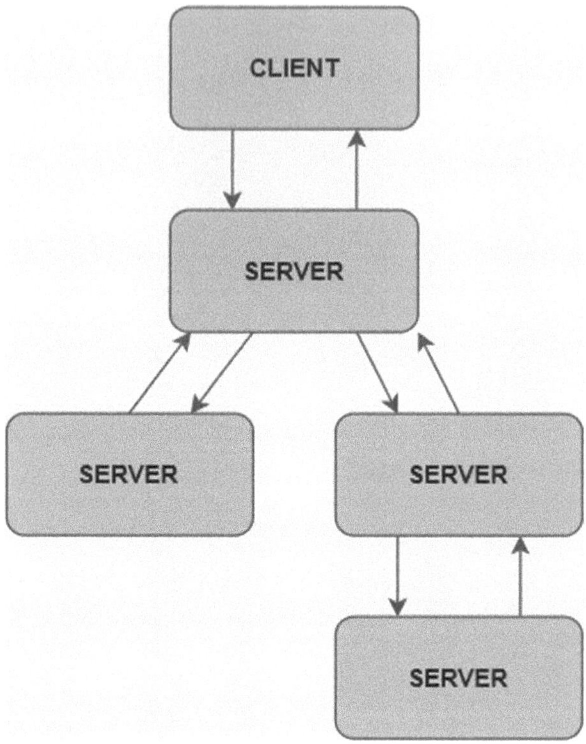

FIGURE 7.4 Multitier architecture.

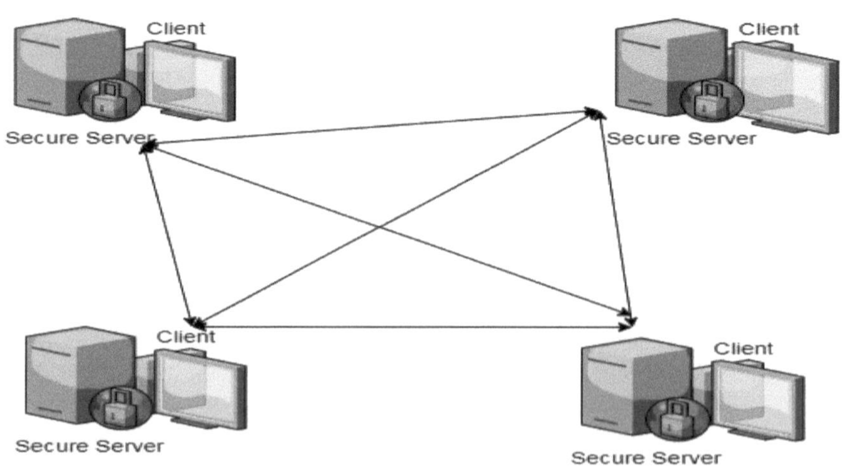

FIGURE 7.5 Peer-to-peer architecture.

which involves clienk and server architectures that was demostrated in above sections.

For example, blockchain technology.

7.2 Semantic distributed architectures

7.2.1 Thematic background

Ontology-based smart grids have grown steadily. Each ontology describes a domain-specific interest that is subjective to the level of granularity. These ontologies are heterogeneous between ontologies that connect under the same domain (Santos, Pinto, & Vale, 2021). In this context, a novel semantic distributed architecture causes significant problems on the semantic web. Semantic web protocol stacks support ontology interoperability to define the relationships between components of smart grids as well as the energy flow in the smart components. Moreover, to build such a semantic distributed architecture, many factors are included to provide reliable communication between the smart components.

Web 3.0 responsible:

- *Semantic Web (SW)*: SW is the next-generation evolution of data generation on the Web. It enhances online technologies by allowing users to create, connect, and share content via search engines. Based on the words rather than keywords or numbers it is able to understand the meaning of key elements.
- *Artificial intelligence*: It is the advancement of intelligent machine combine machine capabilities with NLP, Web 3.0 allows machines to comprehend things in the same way as humans. This results in faster and more relevant capabilities to assist the users, which become increasingly intelligent.
- *3D graphics*: In Web 3.0, 3D design is extensively used in websites and application services. 3D graphics are used in computer gaming, auto engineering, design engineering, e-commerce, and other applications.
- *Connectivity*: In Web 3.0, information is accessible through RDF and RDF schemes through the metadata. It results in the user controlling the data over internet.
- *Ubiquity*: The services can be accessed through multiple architectural information, where every component of a device is connected in the semantic web and whose services are accessed and authorized by the proof and trust layers.

7.2.2 Introduction to semantic web

In the World Wide Web, SW is an extension of new technologies and a set of standards rules that are recommended by the W3C (World Wide Web

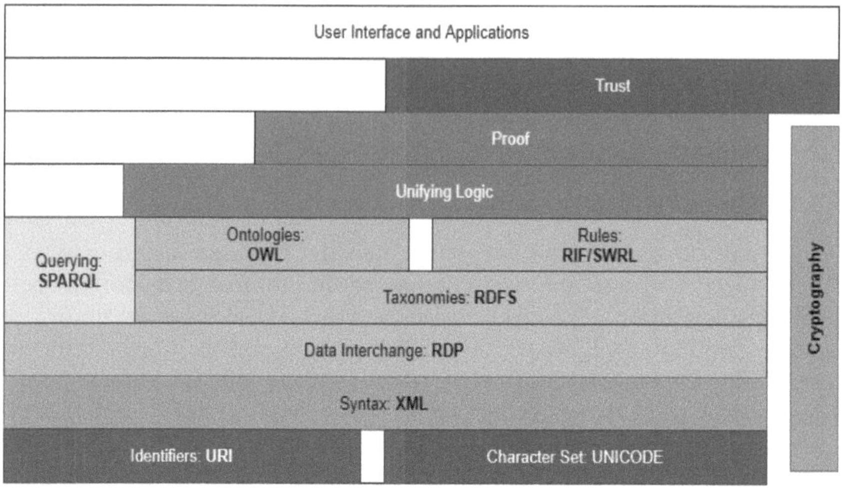

FIGURE 7.6 Semantic web protocol stack.

Consortium) (Krishna, Patruni, & Rao, 2021). In the new evolution, SW represents a major evolution in connecting information through graph structures (Surendra Prasad babu & dasari, 2015). It enables users to link data from one source of knowledge to other sources. The data in ontologies are represented in the form of knowledge bases. In ontologies, knowledge bases are key elements in the retrieval process of the semantic web (Surendra Prasad babu & dasari, 2015). The Fig. 7.6 shows semantic protocol stack with detailed services under different aspects.

In semantic web, there are three technical standards

1. *RDF (resource description framework)*: It a data modeling language for SW that describes objects and their relationships. All SW information is stored in the form of RDFs. There are different standards of RDFs, namely tuple, turtles, etc.
2. *SPARQL protocol and RDF query language*: SPARQL is a query language that retrieves SW information from a repository.
3. *OWL (web ontology language)*: It allows users to specify concepts in a modular manner that can be reused as much as feasible. Each entity is meticulously described and can be selected and constructed in combination with concepts for a multitude of disciplines.

7.2.3 Characteristics of smart grid architecture

- *Participating customers to enable information*: Consumers can demand dependability by modifying how they use and buy electricity. These shifts take place in consumer decisions, which influence buying attitudes and preferences (Zhou, Natarajan, Simmhan, & Prasanna, 2012). New technology, new information on their energy usage, and new price and incentive structures for electricity are among these choices.

- *Generating storage option*: In customer-sited distribution, resources can be supported by a smart grid. Renewable energy, small-scale CHP, and energy storage will all become more integrated throughout the value procession, from supplier to marketers to consumers.
- *Products and services in markets*: When markets are correctly constructed and regulated, consumers can choose from a choice of competing forces. Grid factors that can be actively regulated include energy, capacity, location, time, rate of change, and quality. Markets have the ability to have a big impact on how these variables are managed.
- *Providing power quality*: Not every business, and certainly not every home, necessitates the same amount of power. Power is delivered in a variety of grades at different costs (Daqing & Yinghua, 2015). Control systems keep important agents up to date, allowing for rapid diagnosis and resolution of problems.
- *Optimizing utilization and efficiency*: A smart grid uses cutting-edge technology to get the most out of its resources (Al-Jarrah, Al-Hammadi, Yoo, & Muhaidat, 2017). This allows assets to be operated at higher loads. IT maintenance, which alerts you to the need for apparatus repair at the exact appropriate instance, can help you save money on maintenance.
- *Disturbances, attacks, and natural disasters*: The ability of a system to respond to surprising occurrences by dividing problematic portions while the remainder of the system is functional is referred to as resiliency. Consumers will experience fewer service interruptions as a result of these self-healing efforts, and service providers will be able to better manage their delivery infrastructure.

7.2.4 Semantic smart grid architectures

National Institute of Standards and Technology impacts a conceptual model, which can be used as a reference model for the various segments of the electric system. NIST proposed this model from the perspectives of different roles involved in the smart grids. The conceptual model consists of several domains and their sub-domains (Pranaya, Himarish, Baig, & Ahmed, 2017), each of which contains many actors and applications.

The basic modules of a smart grid:

- *Actor*: One who acts or performs tasks. It may be a device, person, or program. Actor capable of making decisions and capable of communicating with the actor outside the domain through the interface.
- *Application:* Tasks performed by the actor within the domain.
- *Domains*: An area controlled by a particular group.

- *Flow*: It represents the flow of energy or information through the power grid or the communication network.
- *Interface*: It represents the point of access between a system or domain. There exist both electrical and communication interfaces and the interfaces may be bidirectional.

7.3 Smart grids: state-of-the-art

7.3.1 Introduction to smart grids

Traditional power distribution systems transmit electricity from power plants by decreasing the voltage levels. Moreover, power consumption rapidly increases based on device electricity with different voltage levels. In this context, a novel approach of user-friendly featured smart technology delivers the electricity to the end-users by gradually reducing voltage levels (Spanò, Niccolini, Pascoli & Iannacconeluca, 2015). A smart grid is a method that must ensure the safety of users through continuous and reliable communication exchange.

7.3.2 Smart components

The smart grid offers many opportunities for consumers to save energy and for utilities to operate the grid more efficiently and reliably. For instance, some utilities are remote control peak of power demands (Xiao, Rahmani, Li, Gillblad, & Kanter, 2016). There are different components used in smart grid applications. These elements play key roles in the design of smart technologies.

The following components are used in smart grids;

1. *Smart meters*: These elements recognize electrical consumption in various devices. These meters relay information to the central monitoring station. They also recognize power outage notification and quality of power monitoring.
2. *Phasor measurement*: It measures the AC current waveforms and monitors the power quality. There are a wide variety area of measurement systems, namely real-time data, Phasor data connecters, and supervisory control. It sends data to the central monitoring system for battery management.
3. *Information transfer*: Information is transferred in two ways, i.e., ZigBee. It uses IEEE 802.15.4 with low energy that fast start up by using mesh topology.
4. *Distributed generation*: It allows the connection of small-scale renewable generators like solar, wind, fuel cell, and micro turbine generation. They generate energy close to the load so that they work efficiently. The advantage of distributed generation is customer interaction and increased reliability.

7.3.3 Smart grid components

The components in the smart grid play a critical role in the context of distributed systems. Each component is connected with another component concerning the key role of the system (Sidid & Gaur, 2017). The Fig. 7.7 shows, the smart grid architectures in a distributed environment.

1. *Transmission*: The energy source components that transmit the data between the smart agents to other agents in a reliable communication.
2. *Distribution*: The input source energy is transmitted within a distributed environment that should be accessed throughout the network
3. *Customer*: Its agents are customized within the environment which access the components connected in the distributed environment.
4. *Operation*: The functionality accessed through the network.
5. *Service provider*: It allows the customers to access the smart components in service based on available resources.

Smart grids use information technologies to improve the reliability, availability, and efficiency of power systems. In developing the modern SG, applying these technologies to networked applications means that equipment is in the consumer's transmit step by step, making our electrical system sensitive (Spanò, Di Pascoli, & Iannaccone, 2015). In economic and political motivations, the introduction of smart grids also depends on speed and technological advancements.

Potential prospects for smart grids include:

- increased durability
- peak demand drops
- increasing energy efficiency
- improved reliability and network quality
- reduced cost of transmission congestion

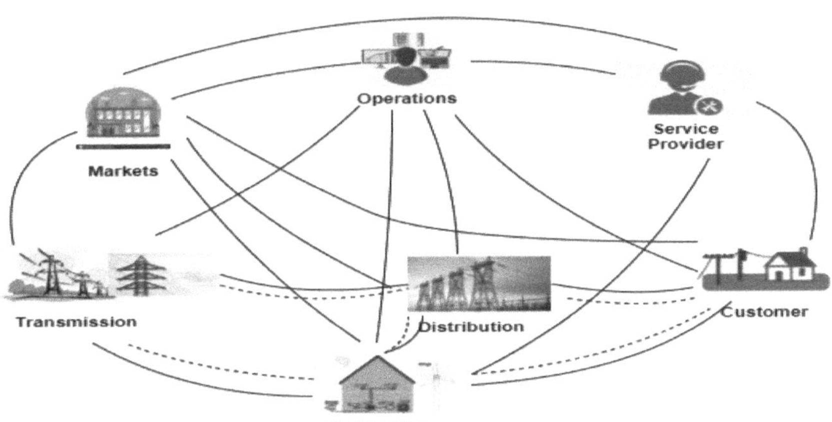

FIGURE 7.7 Smart grid architecture.

Consumer profiles may also change soon. E-vehicles will be a large burden on the network in everyday life. This energy storage will help smooth out the variability in wind and solar resources. By establishing standards between resources and smart grid buyers, resources can be used more efficiently.

7.3.4 Smart distributed architectures

Smart distributed architectural elements and electronics as shown in the Fig. 7.8. This smart distributed device driver is empowered with the operation of the smart grid system and operates to ensure real-time control.

The purpose of SG architecture is to achieve two key goals:

1. Assure grid availability and dependability, autonomous grid monitoring, continuous measurement, and grid control.
2. Also assure energy utilization in distribution and balancing the demand supply.

There are three major prominent layers in a smart distributed architecture:

1. Physical Layer
2. Communication Layer
3. Application Layer

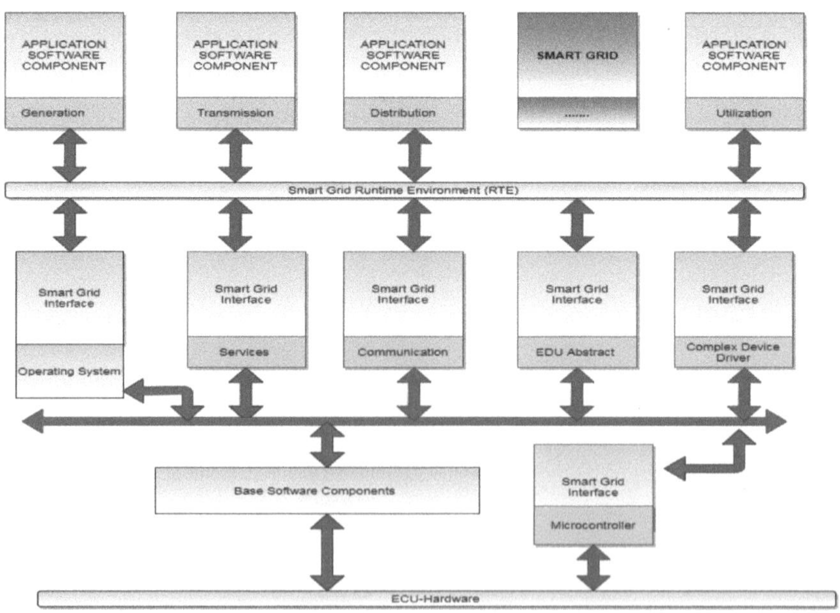

FIGURE 7.8 Smart distributed systems.

Physical Layer: This layer is also known as a base layer. The base layer consists of all the sensor interfaces and device drivers. It has modules that can directly access microcontrollers and peripherals.

Communication Layer: This layer connects to various layers that realize the functionality to abstract details and agreed information exchange protocol from modules that are architecturally placed beneath them. The operating systems (OS) interface, device-independent services, and communication protocol interface are key modules. In a smart grid environment, the customers and energy suppliers are distributively located.

Application Layer: In these application layers, specific features like demand management and load forecast are realized. It consists of sub-layers like service layer and component layer, which implement software functionalities. To end-users, this architecture helps to perform intelligent distribution energy and effectiveness.

Smart grid benefits:

Efficiency: SG benefits distributed generation, smart meters, and energy management systems.

Reliability: It is reliable in wide area monitoring.

Environment: Renewable generation reduces demand on central generation facilities.

7.3.5 Communication exchange

Improvement of the data transmission procedure of the intelligent network includes the parameters that must be into account and formulated. They can ideally recycle the sensor data, and then these procedures implement standard data transfer procedures (Spanò et al., 2015). To reduce the problems in transmission methods for reading the data, the process of reading the smart meter must be confidential and reliable, which includes:

- data transmission media:
  - smart phones
  - remote monitoring
  - central management systems
- smart field applications
  - tracking systems
  - site mapping

7.4 Smart grid technologies

7.4.1 Concept

For a very long time, there has been no change in the actual construction of an electrical system. Encounters have showed in the various levels, midway controlled grids of the 20th century are inappropriate. In the 21st century, electric force network foundation necessary for improved proficiency. Moreover, trustworthiness through robotized control, high power converters, present-day correspondence foundation, detecting and metering advances, and current energy the executives' methods power systems depend on a robust data and correspondence framework. The new innovative network needs an alternate and considerably more intricate one. Its measurement is a lot bigger. This chapter resolves fundamental issues on smart grid advancements fundamentally with data and information communication technology openings and issues. The major goal of this chapter is to give a modern-day take of the present status and quality in smart grid interchanges just as in this field of open exploration issues. It is expected that this chapter will give a superior comprehension of the advancements, expected benefits and exploration tasks of the smart grid and incite curiosity amongst local exploration areas to investigate this favorable exploration region.

7.4.1.1 ZigBee

ZigBee smart energy profile has been acknowledged as the most appropriate correspondence principle for framework private organization area in the US (Lu & Gungor, 2009). Correspondence between smart indicators, just like smart home machines and in-home presentations, is very significant. Like Itron, Elster, and Landis Gyr, numerous AMI sellers favor smart meters that the ZigBee convention can incorporate (Gungor, Sahin, Kocak, & Ergüt, 2011). ZigBee incorporated speaking with integrated gadgets and control them. It gives utility messages to the mortgage holders, and mortgage holders can arrive at their constant energy utilization data.

7.4.1.2 Remote mesh

A cross-section network is an adaptable organization comprising a gathering of hubs, where new seats can join the community and every corner can go about as an autonomous switch. Recuperating the organization's trademark empowers the correspondence signs to discover another course through the dynamic hubs if any hub exits the organization. Particularly in North America, R.F. network-based frameworks are incredibly mainstream. In Smart Measuring Devices, each smart gadget is furnished by a radio segment. Also, in every one of the courses, the metering information is through meters. Each one of the meters shows a repeating sign until the gathered information arrives at the electric organization passage. Then, at that point, the collected information is moved to the benefit through a correspondence organization. A privately-owned business, SkyPilot Networks, utilizes network organizing for smart grid products because of the repetition and elevated accessibility highlights of grid innovation (Gungor et al., 2011).

7.4.1.3 Cell network communication

Existing cell organizations can be a decent choice to convey smart measures and the functionality between many hubs. The current interchange framework dodges utilities from investing functional expenses and extra energy for building a committed correspondence foundation. Cell network arrangements empower smart grid organizations spreading to a comprehensive regional climate for next-generation technologies of cell correspondence advances accessible to utilities for smart metering arrangements. When information moves, the span utility of an enormous amount of information will be produced.

7.4.1.4 Electrical cable communication

Electrical cable correspondence (PLC) is a method that employs the current electrical cables to send fast information to one gadget. PLC has been the best option for correspondence of power meters due to the immediate association with the meter (Vijayapriya & Kothari, 2011), and fruitful executions in metropolitan regions where different arrangements battle to address utility issues. PLC frameworks dependent on the LV conveyance network have been exploration points in China (Laverty, Morrow, Best, & Crossley, 2010). In a normal network, smart meters are associated with the information through electrical cables, and information is moved to a server farm through cell network advances. For instance, any electrical gadget, for example, an electrical cable smart handset-based meter, can be associated with the electrical cable and used to communicate the metering information to a focal area (Gungor et al., 2011).

7.4.1.5 Computerized subscriber lines

Fast computerized information transmission innovation utilizes the ropes of the voice handset organization. It isn't unexpected more noteworthy than 1 MHz through an Advanced Subscriber Lines empowered phone line (Laverty et al., 2010). The existing all-around foundation of Advanced Subscriber Lines lessens establishment costs. Subsequently, numerous organizations picked Advanced Subscriber Lines innovation for this smart framework development. A correspondence frame will one introduced at the client base and the utilization data will be sent over Advanced Subscriber Lines (Gungor et al., 2011). Deutsche Telekom offers numerous administrations in this venture, like perusing utilization information, establishment and activity, information transmission, and so forth. In any case, the throughput of the Advanced Subscriber Lines association relies upon the distance portrayal of the exhibition of Advanced Subscriber Lines innovation (Zhai, 2011).

TABLE 7.1 Industrial applications and services.

Future applications and services	Real-time market
Appropriated ages	Conveyed resources
Grid streamlining	Self-mending grid: flaw security, blackout the executives, dynamic control of voltage, climate information mix, unified capacitor bank control, circulation and substation mechanization, progressed detecting, computerized feeder reconfiguration
Request response	Advanced request upkeep and request reaction, load estimating, and moving
AMI (Advanced metering infrastructure)	Provides far-off meter perusing, burglary identification, client prepays, versatile labor force the board

7.4.2 Applications

The smart grid assumes a significant part in present-day smart advances. Table 7.1 provides the most well-known utilization of Smart grid innovation.

7.4.3 Merits and demerits

7.4.3.1 Merits or benefits of a smart grid

Below are the merits or benefits of a smart grid:

- It decreases power burglary.
- It lessens power misfortunes (transmission, conveyance, and so forth).
- It lowers power cost, meter understanding expense, T&M tasks, and upkeep costs and so forth.
- It lowers hardware disappointments because of programmed activity dependent on fluctuating burden conditions. Request response diminishes weight on resources of smart framework during top conditions, which lessens their likelihood of disappointment.
- It decreases supported blackouts and lessens successively related rebuilding costs.
- It lessens air emanations of CO_2, SO_x, NO_x, and PM-2.5. Thus a smart framework is environment-friendly.
- It lessens oil utilization and wide-scale power outages. Thus a smart grid gives security to individuals by providing constant power.

7.4.3.2 Drawbacks or disadvantages of a smart grid

Below are the demerits or hindrances of a Smart Grid:

- Ceaseless correspondence organization ought to be accessible.
- During crisis circumstances, network blockage or execution are enormous difficulties in a smart grid.
- Cell network suppliers do not ensure administration in unusual circumstances, such as windstorms, substantial downpours, and easing up conditions.
- Some smart grids can be hacked, which can increase or diminish the interest in power. It is costly to introduce a smart meter in contrast to a conventional power meter.

7.4.3.3 Available for smart grid communication technologies

An interchange framework is the critical part of the smart matrix framework (Gungor, Lu, & Hancke, 2010; Sachs, 2010; Spanò et al., 2015). Through coordination of trend-setting innovations and products for accomplishing a more intelligent power grid framework, a gigantic measure of information from various products will be produced for additional examination, control and continuous estimating strategies. Thus it is fundamental for electric utilities to characterize the correspondences necessities and track down the best correspondence frameworks to deal with the yield information and convey a solid, secure, and practical administration throughout the absolute framework. Electric utilities endeavor to stand out enough to interest the smart grid framework and further develop administrations and productivity. Moreover, blackouts after catastrophes in the existing force structure concentrate on the connection between electric networks and correspondence frameworks (Gungor et al., 2010). Various correspondence innovations are upheld by two fundamental correspondences, media and remote, and utilized for information transmission among smart measures and electric services. In few cases, remote correspondences have a few benefits over wired advancements, like minimal expense foundation and simplicity of association with troublesome or inaccessible regions. The idea of the transmission may cause the sign to weaken. Then again, wired arrangements try not to have obstruction issues and their capacities are not subject to batteries, as remote displays are. Essentially, two kinds of data foundations are required for data stream in a smart grid. For the subsequent data stream, cell advances or then again the Internet can be utilized. Critical restricting factors should be considered in the smart metering organization measure, as a season of sending, operating costs, the accessibility of the innovation and country/metropolitan or indoor/outside climate. So, an innovation-decision that fits one environment may not be appropriate for the other. A portion of the smart grid correspondence advancements alongside their benefits and disservices are momentarily clarified in the accompanying chapters. An outline of smart grid correspondence innovations are shown in Table 7.2.

7.4.4 Issues

Great endeavors are being made everywhere to advance clean energy for ecological security and booking energy utilization examples for expanding requests. The network has changed or enhanced to increase control utilization design. The SG will satisfy these prerequisites. The advancements in data and correspondence innovation support the framework. Furthermore, sensor controls upgrade additional support of an existing network to SG. There are numerous difficulties and issues in the execution of the SG. Innovation development has generated colossal advancements in data and correspondence innovation over many previous years. Yet, according to our creative mind, making another correspondence foundation coordinated or entirely based on the Internet is tough from assault and exceptionally solid for making the matrix smarter (Sachs, 2010).

The smart framework will limit making a move all alone by cutting-edge control techniques and choice backings. In nonindustrial nations, utilization of power is just around 25% of world's norm. So, a huge development of force framework network is required to supply energy to all families in a reliable way. The nature is guaranteed to satisfy framework. The current matrix is to be redesigned and extended to ensure excellence stockpiles for families. Again, circulated environmentally friendly power age and alternative to set aside cash by moving burdens from the top burden to an off-maximum period are elevated to lessen supply holes at full hours and pinnacle energy cost (A policy framework for the 21st-century grid, 2010).

The musial data created from the unpredictable force gadgets circuits utilized for mix additionally brings numerous challenges. Another downside of sustainable power age is the prerequisite of the huge land region. The dependability of insurance circuits to disengage it from the current network at whatever point required is an incredible test in incorporating sustainable power sources. The absence of talented labor and incorrect determinations of ideal spots for execution will likewise influence the sustainable power coordination (Indian Smart Grid Forum and Indian Smart Grid Task Force, 2013; Lu & Gungor, 2009). Security and network safety are accomplished distinctly by executing different guidelines (Vijayapriya & Kothari, 2011).

In prior stages, every nation has fostered principles to execute smart matrix thinking about its current norms and simple execution. Universally acknowledged guidelines and safety efforts should be carried out by all nations engaged with SG execution to force a grid with a worldwide matrix (Anees, 2012).

TABLE 7.2 Smart grid communication technologies.

Smart grid communications technologies					
Technology	**Spectrum**	**Data rate**	**Coverage range**	**Applications**	**Limitations**
GSM	900–1800 MHz	Up to 14.4 Kbps	1–10 km	Advanced metering infrastructure (AMI), demand response, HAN	Low data rates
GPRS	900–1800 MHz	Up to 170 Kbps	1–10 km	AMI, demand response, HAN	Low data rates
3G	1.92–1.98 GHz2.11–2.17 GHz (licensed)	384 Kbps-2 Mbps	1–10 km	AMI, demand response, HAN	Costly spectrum fees
WiMAX	2.5, 3.5, 5.8 GHZ	Up to 75 Mbps	10–50 km (LOS)1–5 km (NLOS)	AMI, demand response	Not widespread
PLC	1–30MHZ	2–3 Mbps	1–3 km	AMI, fraud detection	Harsh, noisy channel environment
Zigbee	2.4 GHZ-868–915 MHZ	250 Kbps	30–50 m	AMI, HAN	Low data rate, short range

Fundamental difficulties in SG execution is an absence of purchaser mindfulness of the force area. Buyer support is an unquestionable requirement for the execution of the smart grid to diminish maximum burden utilization and advance appropriated environmentally friendly power age. The smart grid execution will work on the quality and dependability of force supply. It guarantees easy to use and straightforward boundaries for customers with expanded decisions for purchasers, as well as green influence and choice to set aside cash by moving burdens from top periods to off-top periods. However, the shoppers ought to know about innovations and backing the utilities (A policy framework for the 21st-century grid, 2010).

7.5 Industry approaches to smart grids

7.5.1 Introduction to industry approaches of smart grids

A power circulation network that utilizes computerized correspondence intends to identify neighborhood changes in power utilization and respond with no human cooperation is known as a smart grid. The framework conveys power through a 2-way automated resemblance. It permits purchasers to communicate with the matrix. The smart grid conquers numerous downsides of conventional electrical frameworks and gives advantages to customers, power organizations and government foundations. It diminishes energy utilization and decreases the cost to the purchasers by smart grids. Electric stockpile organizations make efficient use of energy and sequentially will want to satisfy the differing load needs of the buyers. The electric network, which is smart in contrast with a customary framework, is known as a smart grid. It distinguishes nearby changes in power utilization and responds consequently without the need for human mediation. It permits two-way correspondence among gride and shoppers. It enables ongoing posts among shoppers and utility to tailor their energy utilization dependent on singular inclinations like cost or potentially ecological concerns. A smart grid is created utilizing current advanced correspondence innovations. Framework speaks with a cell tower to push indispensable boundaries like immediate force utilization, total force use, most significant interest and so forth.

7.5.2 Industry approaches to smart grids

The investment of industrial customers in smart grid framework change is significant due to their utilization impression, hefty energy use, and intricacy in smart energy advances. Dynamic inclusion of modern buyers in smart grid arrangements is essential to guarantee the energy framework change. Although the significance of current customers has been recognized, the observational examinations on the smart matrix still mostly addresses private and business customers. In this manner, in light of four contextual analyses

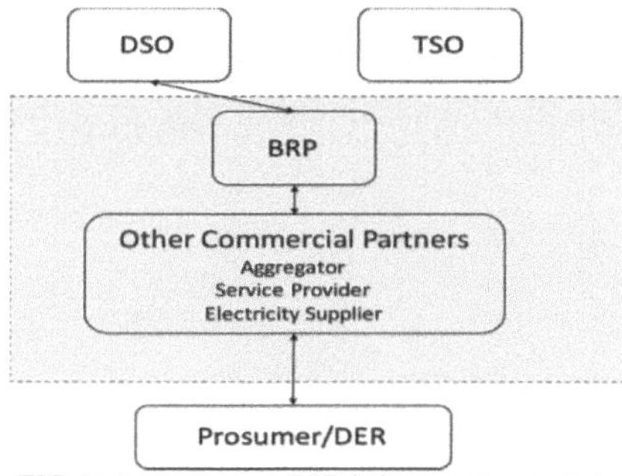

DSO: distribution system operator TSO: transmission system operator BRP: balance response party DER: distributed energy resource

FIGURE 7.9 Predetermined structure to provide elasticity to DSO/grid company for traffic managing. *DSO*, Distributed System Operator.

with two modern shoppers, one energy counseling organization and one power retailer, this chapter examines the variables that impact current customers' acknowledgment of smart grid arrangements and how the powerful factors apply to the smart grid reception stages. Eleven powerful elements have been recognized that affect four phases for industrial buyers' reception of smart grid arrangements (engraving, interpretation, outlining, and adjustment stages). The eleven persuasive elements are: consciousness of numerous unique circumstances, shared help, return-of-venture, usability, adaptability and dynamic estimating, advancement and energy duty structure, client center, arrangement mix, measure improvement, administration quality also (Fig. 7.9) (Table 7.3).

7.5.3 Industry applications

The ascent of new advanced mechanical innovation known as Industry 4.0 has recently acquired a great deal of interest from analysts, producers, and application engineers (Gungor et al., 2010; Sachs, 2010). The goal of Industry 4.0 is to interface and coordinate all objects of the conventional plant world for empowering quicker, more adaptable, and more proficient cycles to deliver greater merchandise at low expenses. Industry 4.0 will bring incredible freedoms for elevating mechanical moves to build fabricating efficiency, shift financial matters, quicker modern development, and combine organic market measures between the processing plants. This will alter the workforce profile, ultimately changing the seriousness of organizations and locales (A policy framework for the 21st-century grid, 2010). In Industry 4.0, remote or wired, associated frameworks situated in various distant spots can communicate with each other utilizing standard web-based conventions and investigating information to anticipate disappointment, arrange themselves, and adjust to changes (Lu & Gungor, 2009). As of now, most gadgets

TABLE 7.3 Participating companies.

Smart grid stakeholder	Role	Current state	Short description
Claus sorensen A/S	Industrial consumer (cold stores/ refrigeration)	Active smart energy participant	Implemented demand response infrastucture
ITW GSE-AXA power	Industrial consumer (manufactures/supplies the ground support products to the aviation industry)	Slightly aware of the Smart energy transition	Plan to change the lighting system in the production facilities
TREFOR group	DSO and electricity retailer	Active smart energy participant	Responsible for grid operation and sells electricity
Udvikling Fyn	Energy consulting	Active smart energy participant	Offers smart energy solutions for the industrial consumers (e.g., light, ventilation)

inside plants are dependent on wired framework working over mechanical conventions; anyway, remote arrangements are progressively assuming a free part to wired arrangements (Indian Smart Grid Forum and Indian Smart Grid Task Force, 2013). To this end, remote mechanical organizations (IWNs) are the key innovation empowering the arrangement of Industry 4.0. They can offer a decrease in energy utilization, influence monetary advantages with the least upkeep and breakdowns, and empower smart creation. However, this does not only influence machine-to-machine correspondence, it will also have extensive ramifications for the interchange of specialists and inserted framework and remote advances (Vijayapriya & Kothari, 2011).

The climate temperature and mugginess checking	In the circulation network climate, counting conveyance bureau, link room temperature, and stickiness sensors are introduced, and screen the climate heat and dampness information. When the temperature and dampness information arrives at the comparing edge, cautioning or alert will be finished.
Water flooding condition checking	Cautioning will be performed when identifying water.
Clamor observing	Introducing clamor sensor, the commotion decibels of appropriation network types of gear (counting segment) observed progressively and warming will be done when the commotion decibel arrives at the relating limit.

Smoke observing	By introducing the commotion sensor in the spot that the smoke will be checked continuously. The sensor will caution the ace station framework when there is apparent smoke produced by seething and consuming.
Pinnacle tendency observing	By introducing the slant sensor in the circulation tower, the slant condition of pinnacle gear will be checked continuously and the information state will be refreshed. The conjecture caution will be incited when the measure of slant scopes to the relating edge.
Dissemination line flaw checking	By introducing the current sensor on the conveyance line, the activity of dissemination lines will be observed progressively. The caution will be completed and the deficiency of branch line will be found when the line is in short-circuiting inconvenience.
Hostile to robbery checking	Observed continuously and the radio recurrence recognizable proof will be done through the on-location activity terminal. As per the typical review time for laborers, a thorough decision whether the entryway is opened unusually will be made and this is as the counter burglary judge for gear.

7.5.4 Opportunities and future challenges

7.5.4.1 Innovation advancements

Arising frameworks make it conceivable to install prepared and advanced correspondences on top of the simple force network. The resultant interchanges foundation fit for more substantial information volumes and dealing with the more prominent information speed. This empowers it to turn out to be further developed in a few straightforward manners:

- A more detectable network. This implies that utilities are coming nearer to where they have total consciousness of matrix state. It likewise means that they can all the more adequately and effectively transport sensor information and control orders.
- A more controllable grid. This is obtained from the new capacity to drive the matrix to any ideal state.
- A more mechanized framework. This furnishes utilities with the capacity to adjust to quickly evolving conditions, and if fundamental, without human mediation.
- A more incorporated framework. Interface siloed utility frameworks and siloed cycles can convey upgraded business benefits.

Utilities currently benefit from developing quantities of private innovation organizations giving progressed correspondence gear for utility prerequisites. These are planned and created to fit with center electrical gear. Such secretly created gear coordinates with the necessary business principles and can be adapted to explicit conditions (e.g., comms-skeptic, contingent upon the particulars of the neighborhood climate).

Much consideration inside the tech business is currently centered around both transmission organizations and substation organizations. These organizations are significant because they dwell inside and among conveyance and transmission substations and utility activity habitats. Such transmission and substation networks should convey a wide range of traffic types for a differed set of capacities, including:

- activity and control interchanges from Supervisory Control and Data Acquisition (SCADA) frameworks, to gather information on substation tasks;
- correspondence among substations and control focuses;
- far-off labor force;
- distant designing access correspondences.

The best substation networks are planned because of these three significant attributes: excess, collective security, versatility to help these numerous correspondence types.

The most developed arrangements accessible to utilities can assist with lessening substation working costs by further developing unwavering quality and improving the joining of circulated energy assets. These new frameworks have some smart capacities which incorporate the accompanying:

- two-way correspondence between transmission-cantered SCADA gear and energy the board frameworks (EMS) for grid observing and control and
- support for deficiency identification, segregation, and rebuilding, just as proactive administration and upkeep of network foundation.

Two sections to the smart framework feel insignificant, although they are not particularly attractive or apparent: networks that associate substations to control focuses and to one another; and networks inside substations. In a smart framework climate, there are undeniably more shared interchanges at all levels than the utility's old star models.

Activity focuses are the main issue for the investigation of functional information finishing into and the smart grid. Applications dwelling in charge habitats play various capacities, going from issue discovery and deficiency segregation to support arranging and planning. Getting the right information from smart matrix smart gadgets to the right applications rapidly and precisely—and reacting in close constant—is the center capacity of the best-planned activity place.

Arising advancements and administrations expect to empower modern information assortment, investigation and reaction answers for conveyance mechanization, and request reaction applications. The central capacities coordinated with the best frameworks incorporate these:

- Virtualized processing for SCADA, AMI conveyance computerization, energizing the executive's applications.

- Application speed increase between activity focuses and server farm.
- Activity focuses security and secure application facilitation.

Smart matrices are quickly advancing toward frameworks that empower circulated knowledge. There is an inactivity pecking order in a smart matrix, with the end goal that a little information should be burned through before it at any point arrives at a control community or server farm.

The utility organization and provincial organizations are the organizations answerable for conveying the information between substation organizations to focal tasks for examination, control, and reaction. To be satisfactory to the difficulties of our occasions, these organizations should provide information in a way that is secure, dependable, elite, and altered—to meet the fluctuating conveyance and the executive's necessities for the various sorts of information traffic streaming over it.

The Field Area Network (FAN) is the "last mile" in the force conveyance framework, which closes with the force purchaser. The FAN associates buyer energy with the executive's frameworks to control utility substations. The FAN empowers deficiency area, seclusion and rebuilding; grid improvement; request reaction; and the board of circulated fuel sources - to give some examples of its crucial capacities. Gadgets associated with the FAN incorporate streamlining hardware (dissemination transformers, capacitor banks, voltage controllers); shortcoming I.D., confinement and reclamation gear (transfers, switches, reclosers and RFIs); and labor force computerization hardware (P.C.s, specific gadget access gear).

The huge vision should be to utilize innovation to change how the world oversees energy and ecological difficulties, empowering start to finish, exceptionally secure worldwide smart matrix. That vision is anticipating a reality where the accompanying components are available:

- Combining utility frameworks and cycles function admirably across all critical network components, determined to expand matrix knowledge and grid efficiencies.
- We are giving arrangements utilizing an exceptionally secure stage that depend on open principles.
- You are conveying expanded grid dependability, security, versatility, and force quality.
- They are empowering smart networks to become stages for advancement.

Making smart framework function admirably relies on perceptibility: quick and solid. And secure trade of information among all parts through inescapable detecting progressively. This is estimated in progressively more limited terms, as short as 120th of a second.

The present electric service organizations have their hands full with heap difficulties. The matrix is old and progressively unreliable, the conventional single direction broadcast model is old-fashioned, market interest designs are changing, and service organizations are under administrative strain to address expanding energy requests significantly more productively. Simultaneously, shoppers are requesting lower energy charges, more substantial assistance, better permeability into their utilization examples, and more decisions about where their energy comes from.

Notwithstanding the geographic area, the methodology service organizations are quickly going to is the "smart grid," a clever interchanges foundation that can effectively coordinate all market interest components associated with the electric matrix. Moving to a smart grid foundation can address a considerable lot of the difficulties that service organizations overall are endeavoring to survive, including:

Wasteful and obsolete grid framework: Most of the world's power framework was constructed when vital energy was somewhat reasonable. The grid was guaranteed by having an abundance age limit in the framework with a unidirectional power stream to shoppers from midway dispatched power plants. Interests in the electric framework were made to satisfy an expanding need, not to change how the framework works essentially.

Erratic fuel sources: When utilizing inexhaustible assets like breeze and sunlight based, power is created just when the breeze is blowing, or the sun is sparkling. Typically, sustainable force creation does not coordinate with the pinnacle request times; all things considered, it offers either excessively or insufficient limits at any one time. Eccentric interest: previously, a request was truly steady and unsurprising. Because of rising energy expenses and developing ecological concerns, clients are changing their use conduct, bringing about more noteworthy interest vacillations. As a result, it has become harder to keep up with unwavering quality and stay away from blackouts in the power framework.

The system for utility guidelines being used today is over 120 years of age. This applied system for utility guidelines being used today is contained in a discourse named "PUBLIC CONTROL AND PRIVATE OPERATION." (This discourse was conveyed by Samuel Insull, the previous colleague to Thomas Edison and the main utility leader of his time.)

The circumstance is magnificent for updating utility guideline structures, particularly considering the changes now in progress. A few governments and a few clients are making the best decision: effectively promising and remunerating utilities that make smart grid speculations. These equivalent utilities move towards more prominent utilization of more productive, more reliable and cleaner fuel sources. That is, truth be told, not just useful for the utility and its clients. It is helping to guarantee a more practical future for us all of us.

1. *Evolving supply*:

   Europe's huge expansion in the sustainable age has been a gift for the worldwide environment. The E.U.'s objective for 2020 is 20% creation from renewables. Fortunately, a few nations are pushing hard. For example, Denmark has effectively worked out in a good way past half of force creation produced by renewables, primarily wind power. Notwithstanding, unusual creation makes discontinuous supplies since most green electrons are delivered just when the breeze is blowing or when the sun is sparkling. Typically, wind power creation does not coordinate well with full interest times. Subsequently, either to an extreme or insufficient limit is in the framework. Also, some sustainable sources feed into a piece of the framework, the medium- and low-voltage matrix, that today is not continually observed (in contrast to the high voltage transmission network). In this manner, lacking data is accessible on the power taken care of into those frameworks.

2. *Changing interest designs*:

   Before, a request was steady and unsurprising. Because of variable energy expenses and developing genuine concerns, clients have been changing their interest conduct. Not exclusively are capricious stockpile sources a test, yet so too are the variances on the interesting side. As a result, it is becoming increasingly intricate to keep up with the power framework's unwavering quality—the critical fixing in staying away from blackouts.

3. *Guideline/compliance*:

   Utilities should look up to the twin predicament: tending to expand energy interest while at the same time lessening their general carbon impression. As governments across the globe have been forcing stricter limitations on ozone harming substance discharges, this makes traditional-age sources like coal more costly. Another model: the US central government presently orders network safety consistency for basic foundation security.

   Numerous administrative bodies are commanding security consistency for basic framework assurance. In the US, NERC-CIP gives explicit necessities that order both physical and online protection arrangements. Purchaser protection and information assurance will likewise drive safety efforts—from homes to the utility's server farms. Furthermore, worldwide substation mechanization norms (as definite in both IEC61850 and IEEE1613) drive innovations inside the actual substation.

4. *New freedoms*:

   With expansion in the number of electric vehicles, utilities will see the rise of an enormous new market. Since supply to the transportation area is as yet ruled today by the oil and gas majors, the electric utilities are quick to track down the most minimal expense and most elevated edge freedoms to override them. One point is to monetarily store power in resources/batteries across the organization to satisfy nearby need floods. This is a huge chance to streamline the power framework, particularly in Denmark, which should adapt to capricious supplies like breeze power creating turbines.

References

Al-Jarrah, O. Y., Al-Hammadi, Y., Yoo, P. D., & Muhaidat, S. (2017). Multi-layered clustering for power consumption profiling in smart grids. *Access IEEE*, *5*, 18459–18468.

Anees, A. S., (2012, December 6–8). Grid integration of renewable energy sources: Challenges, issues and possible solutions. In *Power electronics (IICPE), 2012 IEEE 5th India international conference* (pp. 1–6).

A policy framework for the 21st-century grid. (2010, March). United States agency for international development.

Daqing, X., & Yinghua, H. (2015). An adaptive data management model for smart grid. In *2015 8th international conference on Intelligent Computation Technology and Automation (ICICTA)* (pp. 126–129).

Gungor, V. C., Lu, B., & Hancke, G. P. (2010). Opportunities and challenges of wireless sensor networks in smart grid. *IEEE Transactions on Industrial Electronics*, *57*(10), 3557–3564.

Gungor, V. C., Sahin, D., Kocak, T., & Ergüt, S. (2011, April). *Smart grid communications and networking. Türk Telekom Technical Report-11316-01.*

Indian Smart Grid Forum and Indian Smart Grid Task Force (2013, August 12). *India's smart grid vision and roadmap*. Ministry of Power, Government of India.

Jadhav, R. V., Lokhande, S. S., & Gohokar, V. N. (2016). Energy management system in smart grid using internet of things. In *Power Electronics Intelligent Control and Energy Systems (ICPEICES) IEEE international conference* (pp. 1–4).

Krishna, D., Patruni, M., & Rao, T. (2021). *Emerging social information networks applications and architectures*. doi: 10.1016/B978-0-12-821599-9.00002-9.

Laverty, D. M., Morrow, D. J., Best, R., & Crossley, P. A. (2010, July 25–29). Telecommunications for smart grid: Backhaul solutions for the distribution network. In *IEEE Power and Energy Society general meeting* (pp. 1–6).

Lu, B., & Gungor, V. C. (2009). Online and remote energy monitoring and fault diagnostics for industrial motor systems using wireless sensor networks. *IEEE Transactions on Industrial Electronics*, *56*(11).

Pranaya, Y. C., Himarish, M. N., Baig, M. N., & Ahmed, M. R. (2017). Cognitive architecture based smart grids for smart cities. In *Power Generation Systems and Renewable Energy Technologies (PGSRET) 2017 3rd international conference* (pp. 44–49).

Sachs, G. (2010, May). *A principle-based system architecture fame work applied for defining, modeling & designing next-generation smart grid systems*. Massachusetts Institute of Technology.

Santos, G., Pinto, T., & Vale, Z. (2021). Ontologies to enable interoperability of multi-agent electricity markets simulation and decision support. *Electronics*, *10*, 1270.

Santos, G., Pinto, T., Vale, Z., & Corchado, J. M. (2019). Multi-agent semantic interoperability in complex energy systems simulation and decision support. In *Intelligent System Application to Power Systems (ISAP) 2019 20th international conference* (pp. 1–6).

Sidid, S., & Gaur, S. (2017). Smart grid building automation based on internet of things. In *Power and Advanced Computing Technologies (i-PACT) 2017 Innovations* (pp. 1–4).

Spanò, E., Di Pascoli, S., & Iannaccone, G. (2015). Internet-of-things infrastructure as a platform for distributed measurement applications. In *Instrumentation and Measurement Technology Conference (I2MTC) 2015 IEEE International* (pp. 1927–1932).

Spanò, E., Niccolini, L., Pascoli, S. D., & Iannacconeluca, G. (2015). Last-meter smart grid embedded in an internet-of-things platform. *IEEE Transactions on Smart Grid*, *6*(1), 468–476. Available from https://doi.org/10.1109/TSG.2014.2342796.

Surendra Prasad Babu, M., & Dasari, S. K. (2015). Design and development of semantic web based guava expert system for disease classification. *International Journal of Advanced Research in Computer and Communication Engineering*, *4*(9).

Vijayapriya, T., & Kothari, D. P. (2011). Smart grid: An overview. *Smart Grid and Renewable Energy*, *2*, 305–311, Nov.

Xiao, B., Rahmani, R., Li, Y., Gillblad, D., & Kanter, T. (2016). Intelligent data-intensive IoT: A survey. In *Computer and Communications (ICCC) 2016 2nd IEEE international conference* (pp. 2362–2368).

Yun, Z., Junjie, L., Ji, C., & Hua, W. (2017). A framework research of power distribution equipment condition monitoring cloud platform based on RESTful web service. In *Energy Internet and Energy System Integration (EI2) 2017 IEEE conference* (pp. 1–6).

Zhai, M. Y. (2011). Transmission characteristics of low-voltage distribution networks in China under the smart grids environment. *IEEE Transaction on Power Delivery*, *26*(1), 173–180.

Zhou, Q., Natarajan, S., Simmhan, Y., & Prasanna, V. (2012). Semantic information modeling for emerging applications in smart grid. In *2012 ninth international conference on information technology—New generations* (pp. 775–782). https://doi.org/10.1109/ITNG.2012.150.

Chapter 8

Application of artificial intelligence and information and communication technology in the grid agricultural industry: business motivation, analytical tools, and challenges

Adedoyin A. Hussain[1,2], Barakat A. Dawood[1,2], Chadi Altrjman[3], Sinem Alturjman[2,4] and Fadi Al-Turjman[2]

[1]*Computer Engineering Department, Near East University, Mersin, Turkey,* [2]*Department of Artificial Intelligence Engineering, Research Center for AI and IoT, AI and Robotics Institute, Near East University, Mersin, Turkey,* [3]*Department of Chemical Engineering, Waterloo University, Waterloo, Canada,* [4]*Department of Artificial Intelligence Engineering, Near East University, Mersin, Turkey*

8.1 Introduction

Recently, agriculture has developed in preparing, promoting, and the dispersion of yields and livestock items. Farming is the bedrock of the manageability of any economy (Kekane, 2013). The use of AI with horticulture will be empowered by other innovative advances, including huge information investigation, advanced mechanics, the web of things, the accessibility of modest sensors and cameras, drone innovation, and even wide-scale web inclusion on geologically scattered fields as depicted in Fig. 8.1. It has a vital impact on long-haul monetary development and underlying change (Johnston & Kilby, 1975; Kuznets, 1973; Syrquin, 1988); however, may differ by nation (Dekle & Vandenbroucke, 2012). With the worldwide population increases, it is basic that horticultural practices are evaluated with the point of proffering imaginative ways in supporting and improving agrarian exercises. Before, rural exercises were restricted to food and yield creation (Fan et al., 2012). By investigating soil, for example, temperature, climate, soil examination, dampness, and memorable yield execution, AI frameworks will have

Sustainable Networks in Smart Grid. DOI: https://doi.org/10.1016/B978-0-323-85626-3.00002-8

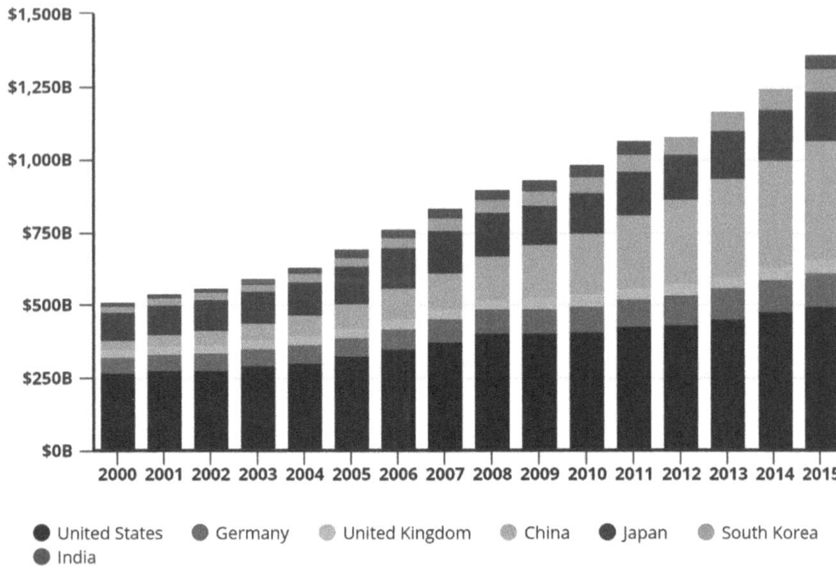

FIGURE 8.1 Continental information and communication technology and artificial intelligence trends (Henry-Nickie, Frimpong, & SunFriday, 2019).

the option to give prescient experiences of the yield and what are the ideal dates to plant and reap in a particular territory, hence improving harvest yields and reducing the utilization of water, composts, and pesticides. Right now, rural exercises fill in as the essential wellspring of occupation, improving gross domestic product (GDP) (Oyakhilomen & Zibah, 2014), being a wellspring of public exchange, decreasing joblessness, giving crude materials to creation in different ventures, and generally improving the economy (Awokuse, 2009; Badiene, 2008; Block & Timmer, 1994). Utilizing AI innovations cause the effects on common environments to be discounted, and laborer security may expand, which will keep food costs down and guarantee that the food creation will continue to increase with the expanding populace. The combination of AI with agribusiness will be empowered by other innovative advances, including enormous information investigation, mechanical technology, the web of things, the accessibility of modest sensors and cameras, drone innovation, and even wide-scale web inclusion on topographically scattered fields. Until this point, AI has been utilized in horticulture in piecemeal methodologies, being treated as an umbrella of computational procedures used to improve execution, for instance, measure control, asset usage, and making decisions. In any case, to genuinely outfit the intensity of AI in horticulture requires coordination of assorted AI apparatuses across the aggregate of the creative framework and appropriation of improved frameworks-based methodologies (National Academies of Sciences Engineering Medicine, 2019). A portion of the methods inside AI that are being utilized for horticulture incorporate machine learning, vision, and the Internet of Things (IoT). The resultant blended utilization of AI procedures can give improved proficiency of information assortment, investigation, and undertaking execution bringing about a more impressive knowledgebase than workable for any individual segment. The specialists in the framework can be extensively sorted into three classes namely software agent (SA), human agent (HA), and actual specialists (PA). To successfully adjust AI for rural creation frameworks, they should accordingly be seen as multi-specialist frameworks made out of heterogeneous specialists whose practices, mentalities, and cooperation decide the status of the framework and whose activities are made out of a blend of these procedures. Treating the utilization of AI in agribusiness in an arrangement of frameworks (SoS) system can help in tending to basic regions and requirements that horticulture is looking for in versatile and supportable creation as portrayed in Fig. 8.2.

As a result of the incredible capability of ICTs, a few researchers have wondered about the digital agriculture revolution (DAR), and they trust it can assist with tackling a few difficulties and this area is confronted (Rotz et al., 2019). While the DAR can be required to be practically identical in size with the green revolution (World Bank, 2017), the advancement worldview brought by the DAR is unique. During green unrest, advances were pointed toward modifying the agroecosystem through manures, pesticides, and hereditary qualities. In a like manner, all advanced advances in farming have one basic component, which is in the utilization

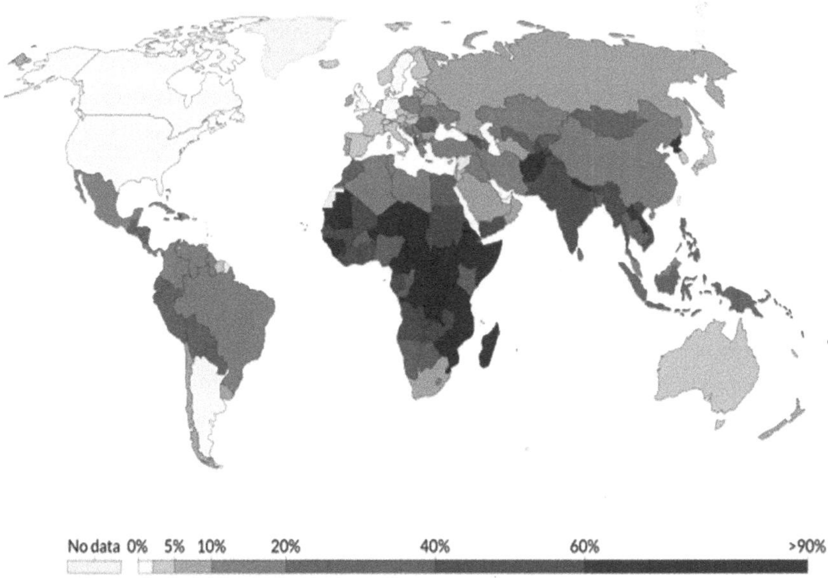

FIGURE 8.2 Global agricultural trends on a global scale (Roser, 2013).

of information. The DAR is changing the choice climate through data arrangement. With it, new stages produce data pointed toward supporting choices by bringing down vulnerability. In this specific situation, water systems are one of the key areas where ICT-data would have the main applications with the most noteworthy advantages (Baseca, Sendra, Lloret, & Tomas, 2019). Choice cycles would now be able to move from prudent and wasteful decisions constrained by vulnerability to choices dependent on sound data. Numerous ICTs have been created in the water system (Jeuland, Hansen, Doherty, Eastman, & Tchamkina, 2019; Masseroni, Arbat, & De Lima, 2020). This articulation features exactly how significant the phase of rural advancement we are living in as per researchers. This area is more vulnerable than others, and ICT has an incredible potential to decrease vulnerability and help to confront the difficulties presented by clashing uses, water shortage, and extraordinary climate occasions (Cavazza, Galioto, Raggi, & Viaggi, 2018; Clara Eli-Chukwu, 2019; Galioto, Raggi, & Viaggi, 2017; O'Mahony, Dalrymple, Levin, & Greenland, 2016).

8.1.1 Comparison to other surveys

In the mechanical and utility areas, the appropriation of ICT to help choices on water use and portion is quickly filling in what is distinguished by the International Water Association as the computerized water venture (Ceo, Foundry, & Webb, 2019). Here, elements for ICT usage are very intricate and the absence of foundations, absence of supporting advancements, monetary limitations, and information gaps regularly present problems to ICT dissemination (Molina-Maturano, Speelman, & De Steur, 2020). In flooded farming, the advanced water venture is more troublesome because of the inborn qualities of the area. For instance, the low precision of accessible gadgets makes numerous stages pointless to help ranchers' water system choices (Galioto et al., 2017). All the more inconsequentially, in certain events, water authorities (WAs) and ranchers may choose to not execute an ICT because the data passed on by the ICT is not full-grown enough or reasonable for which it is utilized (Galioto, Chatzinikolaou, Raggi, & Viaggi, 2020).

The low precision of the data given through existing ICT gadgets is likewise recorded at the degree of WAs for the designation of water assets through outside channels (Cavazza et al., 2018). Accordingly, the computerized progress for water system cannot be naturally cultivated by the area. In any event, when the data passed on through ICT is adequate enough to produce obvious advantages, conduct obstructions can prevent the digitalization cycle (Kirchhoff, Lemos, & Engle, 2013). These conditions feature the need to comprehend choice cycles and to plan arrangements to support ICT improvement and take-up for water the board in horticulture (Irrigants d'Europe Water Framework Directive Position Paper, 2018; Giupponi, 2014). Limitations to digitalization will not just slow ICT execution, yet they likewise hazard bargaining the advancement of ICT developments in agribusiness, leaving the area without date apparatuses and not successful to confront issues of water shortage and clashing employments. In the applied financial matters, a few investigations are tending to the subject of ICT execution in horticulture and water the board. Even though researchers concede to the hypothetical settings where ICT data is significant, experimental applications show discordances and ICT benefits are still hazy. The main works in applied financial aspects gauge the advantages of ICT usage by characterizing the conditions in which data has an incentive for a DM. Further, there are holes in the displaying of choice on ICT execution to represent the vulnerability settings that influence DM's conduct and may block ICT take-up. These can be brought about by the accompanying issues which are the ICT acquires defective data which isn't 100% dependable and the DM does not have the foggiest idea about ICT's unwavering quality. Appropriately, there are various wellsprings of vulnerability around the choice for ICT execution. This doesn't permit us to see how discernments on ICT unwavering quality influence the rancher's or WA's conduct and, thusly, the choice on ICT execution. Nonetheless, vulnerability is regularly demonstrated not recognizing its various sources, and issues created by an absence of information on ICT unwavering quality are ignored. The oddity of this examination is in the utilization of AI and ICT hypotheses to give the premise expected to show horticultural administration. In Table 8.1 we show the correlation with other important investigations. In the greater part of applied examinations on AI and ICT selection in the agribusiness industry, hazard is demonstrated in the choice cycle to be the solitary component that shapes vulnerability. Conflictingly to these examinations, we present the idea of vagueness clarifying the portion of AI and ICT in this industry.

8.1.2 Scope of the survey

This overview audits past investigations on the utilization of AI and ICT in the agricultural framework. It assesses how smart frameworks can be sent in the development advantage and how these innovations have been used for helpful horticulture. The part diagrams have completely disclosed the essential ways to deal with yield development. This part looks at the different brilliant advancements that can be utilized to yield agrarian exercises and their correlation for speedy and better arrangement. This is the place where AI and ICT assumes a significant job. Agribusiness involves a lot of decisions and vulnerabilities. Ranchers should adapt to these vulnerabilities. From season to prepare the climate shifts, the costs of cultivating materials vacillate, soil debases, crops are not practical, weeds choke out yields, bugs harm crops, and the atmosphere changes.

TABLE 8.1 Comparison to other research.

Reference	Efficiency	Artificial intelligence	Information and communication technology	Robustness	Scalability
(Ceo, Foundry, & Webb, 2019)	✓	✓		✓	
(Molina-Maturano, Speelman, & De Steur, 2020)			✓		
(Galioto, Chatzinikolaou, Raggi, & Viaggi, 2020)		✓		✓	✓
(Kirchhoff, Lemos, & Engle, 2013)	✓	✓	✓	✓	
(Irrigants d'Europe Water Framework Directive Position Paper, 2018)		✓	✓	✓	
(Giupponi, 2014)		✓	✓	✓	
(Gupta, 2019)	✓	✓		✓	
(Davenport & Ronanki, 2018)	✓	✓		✓	
Our survey	✓	✓	✓	✓	✓

In like way, our commitment can be depicted out as follows:

- This overview outlines a wide and fruitful review of AI and ICT in the farming business.
- We put forward reasonable courses of action using these methodologies in farming.
- The motivations of utilizing these advances in the agrarian business are advanced.
- Usage of AI and computer networks are put forth.
- The issues and troubles related to these innovations are advanced.
- Finally, we see open issues and challenges.

The review is formed as follows. In Section 2, we present critical data on AI in the agricultural business and show basic motivations driving the usage of this innovation. In Section 3, we portray the use of ICT in the agricultural business and its noteworthiness. At that point, harnessing ICT and AI in the agricultural business is shown in Section 4. The usage of AI and computer networks in the agricultural industry and the effect of ICT and AI are analyzed and examined in segments 5, and 6, appropriately. Section 7 highlights issues and proposals acquired from this paper including the challenges that will be clarified later on. Ultimately, Section 8 wraps up this paper with the conclusions. Table 8.2 gives a rundown of the used abbreviations and their importance.

TABLE 8.2 Used abbreviations.

Terms	Meaning
PC	Personal computer
DAR	Digital agriculture revolution
NN	Neural networks
ICT	Information and communication technology
ML	Machine learning
IoT	Internet of things
ANN	Artificial neural network
CNN	Convolutional neural system
DL	Deep learning
ROI	Return on investment
AI	Artificial intelligence
HONN	Higher-order neural network
CA	Computer audition
DEM	Digital elevation model

8.2 Application of artificial intelligence in the agricultural industry

AI innovation supports various areas to help profitability and effectiveness. Similarly. computer-based intelligence in horticulture is assisting ranchers with improving their effectiveness and decrease ecological unfriendly effects. Simulated intelligence arrangements are helping to beat the conventional difficulties in each field. The agribusiness business firmly and transparently grasped AI to change the general result. Adjusting AI innovation assists with control and deals with any excluded regular condition. With this, the advantage of AI in rural industry is depicted in Fig. 8.3. Man-made intelligence is moving how our food is delivered where the rural area's outflows have diminished by 20% (Gupta, 2019).

Today, most new companies in horticulture are adjusting AI-empowered ways to deal with increment the proficiency of farming creation. Actualizing AI-enabled approaches could recognize infections or atmosphere changes sooner and react keenly (Davenport & Ronanki, 2018). The Market study report expressed that the worldwide Artificial Intelligence (AI) in Agriculture market size is required to arrive at 1550 million in the coming years. The organizations in horticulture with the assistance of AI are handling the farming information to diminish the unfriendly results. The best piece of actualizing AI in agribusiness that it will not dispense with the positions of human ranchers rather it will improve their cycles. In this subsegment, we will portray zones in which AI can help advance the rural business.

- AI arrangements can tackle the difficulties ranchers face, for example, atmosphere variety, an invasion of bugs and weeds that decreases yields.
- AI usage accentuation on checking faulty harvests and improving the potential for sound yield creation.

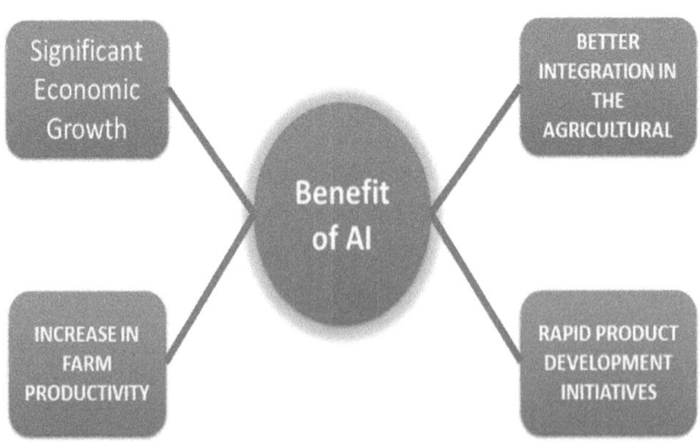

FIGURE 8.3 Illustration of the benefit of artificial intelligence in the agricultural industry.

- AI gives more proficient approaches to deliver, collect and sell fundamental yields.
- AI is being utilized in applications, for example, mechanized machine changes for climate determining and infection or vermin distinguishing proof. The development in Artificial Intelligence innovation has reinforced agro-based organizations to run all the more productively.
- Artificial knowledge can improve crop the board rehearses hence, helping numerous tech organizations put resources into calculations that are getting valuable in horticulture.

8.2.1 Management of disease

Animals and plant sicknesses are a significant restricting component concerning the expansion of yield. To have an ideal yield in rural areas, infection prevention is important. A few variables assume job in the hatching of these sicknesses which assault plants and creatures, which incorporate hereditary, soil type, downpour, dry climate, wind, temperature, and so on. Table 8.3 records the AI applications in this area. Because of these variables and the temperamental idea of certain infections, dealing with the impacts is

TABLE 8.3 Summary of the management of diseases.

References	Technique	Limitation	Strength
(Weed Science Society of America, Facts)	Computer vision system, genetic algorithm, Artificial neural network	Dimension-based detection which may affect good species.	Works at a high speed. Can multitask.
(Balleda, Satyanvesh, Sampath, Varma, & Baruah, 2014)	Rule-based expert, data base (DB)	Inefficacy of DB when implementing in large scale.	Accurate results in the tested environment.
(Munirah, Rozlini, & Siti, 2013)	Expert system using rule-base in disease detection	Time consuming. Needs constant monitoring to check if pest has built immunity to the preventive measure.	Faster treatment as diseases are diagnosed faster. Cost effective based on its preventive approach.
(Kolhe, Kamal, Saini, & Gupta, 2011)	FL & text-to-speech converter	Requires high speed internet. Uses a voice service as its multimedia interface.	Resolves plant pathological problems quickly.

a major test, particularly in huge scope cultivating. To accomplish these is tedious and not in the least that practical (Weed Science Society of America, Facts), thus the requirement for utilization of AI approach for infectious prevention. To viably control sicknesses and limit misfortunes, a rancher ought to receive an incorporated infectious prevention and the board model that incorporates physical, compound, and organic measures (BEA, 2018). Explanation block gives of the rationale followed by the piece of the master framework (Balleda et al., 2014). A text-to-speech (TTS) converter is utilized for giving the ability of text-to-talking user interface (UI).

8.2.2 Management of crops

The application of crops management is summed up in Table 8.4. Harvest is a comprehension of the class indicated by their planning and flourishing soil type which will surely build crop yield. It is summarized as the exercises that improve the development and yield of rural items. Precision crop management (PCM) is a rural administration framework intended to target yield and soil contributions as per field necessities to upgrade benefit and secure the climate. PCM has been hampered by the absence of ideal, appropriated data on harvest and soil conditions (Moran, Inoue, & Barnes, 1997). Ranchers should consolidate different harvests to adapt to water shortfall coming about because of soil, climate, or restricted water system. Timing, force, and consistency of dry season are significant highlights for picking among trimming choices (Debaeke & Aboudrare, 2004). PROLOG uses climate information, apparatus limits, work accessibility, and data on reasonable and organized administrators, farm haulers, and executes for assessing

TABLE 8.4 Summary of the management of Ceops.

References	Technique	Limitation	Strength
(Lal et al., 1992)	PROLOG	Location-specific.	Removes less used farm tools from the farm.
(Snehal & Sandeep, 2014)	ANN	Only captures weather as a factor for crop yield.	Predicts crop yield.
(Henten et al., 2002)	ROBOTICS	Slow picking speed and accuracy.	Has 80% success rate in harvesting crops
(Pilarski et al., 2002)	ROBOTICS Demeter	Expensive: Uses a lot of fuel.	Can harvest up to 40 hectares of crop
(Ji et al., 2007)	ANN	Time-consuming, limited to a particular climate.	Can accurately predict rice yield.

the operational conduct of a ranch framework. Appropriate comprehension of climate designs helps in the dynamic cycle that will bring about high and quality harvest yield (Aubry, Papy, & Capillon, 1998). It additionally assesses crop creation, net income, and net benefit for singular fields and the entire ranch (Lal, Jones, Peart, & Shoup, 1992). Boundaries like soil type, pH, phosphate concentration, natural carbon, magnesium, manganese, iron, temperature, and dampness (Snehal & Sandeep, 2014). The harvest forecast technique is utilized to foresee the appropriate yield by detecting different soil boundaries and boundaries identified with the air. Demeter is a PC-controlled speed-paddling machine, outfitted with a couple of camcorders and a worldwide situating sensor for route. The utilization of AI in reaping cucumber contains the individual equipment and programming parts of the robot including the independent vehicle, the controller, the end-effector, the two PC vision frameworks for identification and 3D imaging of the products of the soil climate and, at long last, a control plot that creates impact-free movements for the controller during collecting (Henten et al., 2002). It is equipped for arranging tasks for a whole field, and afterward executing its arrangement by cutting harvest lines, going to cut progressive lines, repositioning itself in the field, and identifying unforeseen snags (Pilarski et al., 2002). More modest informational collections require fewer concealed hubs and lower learning rates in model enhancement (Ji, Sun, Yang, & Wan, 2007). Field-explicit precipitation information and climate factors can be utilized for every area. Changing ANN boundaries influence the precision of rice yield expectations.

Soil management

Sound information on different soil types and conditions will improve crop yield and ration soil assets. Soil is a basic piece of farming exercises. The utilization of fertilizer and excrement improves soil porosity and collection. Metropolitan soils may contain poisons that can be examined with a conventional soil review approach (Kimpe & Morel, 2000). A superior accumulation demonstrates the expansion of natural materials that assume a significant part in forestalling soil outside layer arrangement. The utilization of natural materials is fundamental to improving soil quality (Pagliai, Vignozzi, & Pellegrini, 2004). The creation of vegetables and other palatable harvests is regularly fundamentally influenced by a few soil-borne microbes that require control through soil (Abawi & Widmer, 2000). The Management-oriented modeling (MOM) limits nitrate draining as it comprises of a bunch of created conceivable administration choices, a test system that assesses every other option, and an evaluator that figures out which elective meets the client weighted various standards. Affectability to soil corruption is verifiable in the evaluation of the maintainability of lands, with acknowledgment of the way that dirts shift in their capacity to oppose change and recuperate (Syers, 1997). The application of AI is summed up in Table 8.5. MOM utilizes hillclimbing as an essential pursuit technique that utilizes a strategic inquiry strategy to locate the most limited way from start hubs to objectives (Li & Yost, 2000). It is the utilization of tasks, practices, and medicines to improve soil execution. Information on designing for developing the Soil Risk Characterization Decision Support System includes three phases which are calculated plan, information procurement, and framework execution (Lopez, Garcia, Schuhmacher, & Domingo, 2008). It is conceivable to receive elective culturing frameworks to forestall soil actual debasement. The elements of soil dampness are described and assessed by a far-off detecting gadget implanted in a higher-order neural network (Elshorbagy & Parasuraman, 2008). An ANN model predicts soil surface in light of qualities got from existing coarse goal soil maps joined with hydrographic boundaries gotten from a digital elevation model (Zhao et al., 2009).

8.3 Application of information and communication technology in the agricultural industry

The subject of ICT selection in horticulture is of developing pertinence and various ICT advancement activities have been completed to help the area (Aker, Ghosh, & Burrell, 2016). This may be affirmed by the developing assortment of articles distributed on the point and by the interest brought up in the applied financial writing. The utilization of such innovations for water systems are viewed as perhaps the most encouraging applications, even though their application is still dominatingly spearheading. In Fig. 8.4 we

TABLE 8.5 Summary of the management of soil.

References	Technique	Limitation	Strength
(Li & Yost, 2000)	MOM	Takes time. Limited only to nitrogen.	Minimizes nitrate leaching, maximizes production.
(Lopez et al., 2008)	Fuzzy Logic:SRC-DSS	Needs big data. Only a few cases were studied.	Can classify soil according to associated risks.
(Elshorbagy & Parasuraman, 2008)	ANN	It does not improve soil texture or proffers solution to bad soil texture.	Successfully reports soil texture.
(Zhao et al., 2009)	ANN	The prediction will fail with time as weather conditions are hardly predictable.	Able to predict soil moisture.

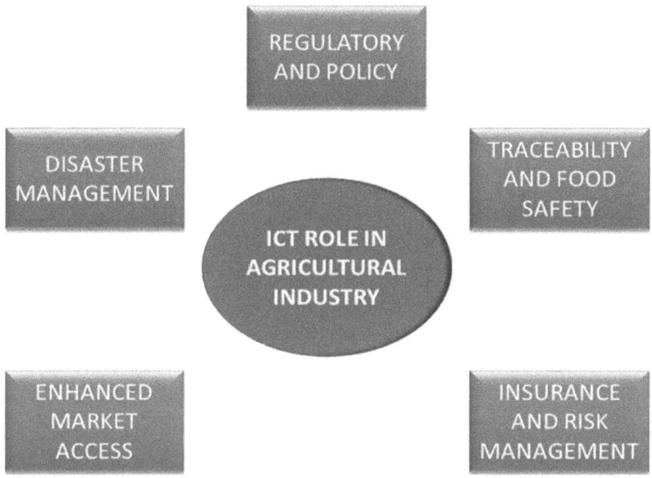

FIGURE 8.4 Illustration of the benefit of information and communication technology in the agricultural industry.

show the advantage of ICT in the farming business. ICT can give benefits just if the conveyed data is ultimately executed by DMs to improve choices (Vogel, Letson, & Herrick, 2017). Despite the extraordinary interest that appeared by researchers for ICT, as a rule, their effect is not all-around characterized and aftereffects of applied financial investigations are very factor. It isn't just data convenience which conditions ICT-benefits. DM's can emphatically influence ICT-data usage as well. This happens just when the data content answers DMs' necessities (Furman, Roncoli, Crane, & Hoogenboom, 2011) and when how data is passed on, for example, idealness of data arrangement or spatial scale permits the DM to actualize it. Authors in (Tumbo et al., 2018) in their investigations found that ranchers in Tanzania are looking for ICT-data to adjust to CC, particularly in their water system exercises. Authors in (Nesheim, Barkved, & Bharti, 2017) found that the utilization of ICT in India has not arrived at its latent capacity and numerous ranchers have not utilized it. However, they feature that vulnerability on ICT unwavering quality cut-off point's data take-up. This is basically because the ranchers do not understand or have questions on ICT dependability (Nesheim et al., 2017). They found that when DMs saw data as dependable, this aided data take-up and proficient water the board. Authors in (Kirchhoff et al., 2013) completed subjective and quantitative investigations on ICT selection for water executives in the United States and Brazil. Authors in (Haworth et al., 2018) completed a survey on ICT improvement activities across the world. They recognized various hindrances that contradict ICT-data usage, among which the most significant ones are related to data quality and saw dependability. Their discoveries underline that between the 27 ICT just a few were utilized by the tended to ranchers. It is unsurprising that ICT or the data itself are viewed as problematic or are deciphered, and DMs do not understand the message or acquire no additional profit. Authors in (O'Mahony et al., 2016) indicated that the absence of information on the innovation's unwavering quality can even forestall accomplishing reasonable in Australian farming. Such discoveries are affirmed by a few quantitative investigations, out of which a choice of observational application for ICT benefits assessments as appeared in Fig. 8.4. To acquire better bits of knowledge from this point, it would be ideal if you allude to crafted by authors in (Meza, Hansen, & Osgood, 2008) which give individually surveys of studies to ICT-data benefits assessment in water and more extensive agrarian area. Even though it is not extensive of the studies done in the field, it is sufficient to affirm the irregularity of the effects brought by the execution of existing ICT in horticulture.

8.3.1 Improves economic growth

This infers that utilization of ICT in horticulture may improve public monetary development by encouraging cultivating exercises inside the required period. Data and Communication Technology apparatuses could be embraced in the Agricultural area to quicken the turn of events and may consequently prompt development in the public economy. Similarly, the discoveries of authors in (Behera et al., 2015), indicated that e-horticulture enhances the lives of ranchers and end-clients in a practical advancement through e-administration, information the board entries, e-booths, and basic assistance habitats at grass root level. The above discoveries were upheld by authors in (Al-Turjman, 2019) that ICT makes mindfulness among ranchers and craftsmen through Geographical Information System (GIS) for legitimate arranging and the board of common assets. It likewise gives the premise to the worldwide network to screen improvement and approval of calculated models and systems, to bundle and disperse them once tried. The creators reasoned that the job of IT to create e-horticulture and prominence of life in the provincial territory is grounded. The exploration of authors in (Pande & Deshmukh, 2015) depicted the importance of ICT in horticulture yet neglected to address satisfactory examination techniques. A normal Indian rancher could get proper data concerning agro-inputs, crop creation advances, agro preparing, market backing, and agro finance through IT.

8.3.2 Agricultural chances

Similarly, authors in (Munyua & Adera, 2009) found that cutting-edge ICTs can improve farming profitability by conveying information and data to provincial agrarian networks. Using ICT, business people can make new open doors by entering the global market and get contact with new accomplices, and trade pertinent data for their business supportability. These have been affirmed by authors in (Singh, Kumar, & Singh, 2015), that ICT facilitates agricultural productivity. This contention relates with authors in (Verdouw, Robbemond, & Wolfert, 2015) when ERP framework information are naturally partaken in the total framework after information passage. Although this information might be sufficient, the investigation is lacking for agricultural nations. Essential information utilized through field visits to establishments in Ghana and Uganda. Hence, more examination needs to be led in other agricultural nations to prove the contention.

8.3.3 Profit ascent

Data and Communication Technology assumes a significant job in upgrading the effect and execution of horticultural creation and likewise, straightforwardly ease neediness through credit and monetary administrations (Ramli et al., 2015). Furthermore, it limits immediate and backhanded costs, especially ad costs, and simultaneously improve business measure. Business visionaries can expand creation and offer more items to possibilities purchasers and improves viable administration of deals. All the more, ranchers could get to applicable data through ICT anytime, and this empowers them to make occupations with improvement organizations and different ranchers, and at last yield higher rural profitability (Abdullah & Samah, 2013). This could be accomplished utilizing sites to furnish ranchers with the capacity to trade data with different ranchers and expansion officials which are modest contrasted with phone activity. These statements were upheld that ICT gives convenient data on what, when, where, why, and how to create and sell horticultural items. These discoveries are predictable with authors in (Hussain, Bouachir, Al-Turjman, & Aloqaily, 2020) who discovered that there is the probability of decreases in business cost with the utilization of ICT. Moreover, ICT-based market data frameworks have a demonstrated history for improving the provincial jobs in center pay workers of agricultural nations where they have been presented. Since it limits the cost of data search and permits more ranchers to contribute to business agribusiness and support cultivating movement through lightening destitution among ranchers in non-industrial nations. It has been sketched out that a combination of overall worth chain administration concerning web development in food store network gives a chance to the ranchers to receive reward on global worth chain consideration by bringing down the level of anomaly. These outcomes have been upheld by authors in (Singh, Sankhwar, & Pandey, 2015) in their magnificent audit that plans to examine the likely advantage of Indonesia cocoa ranchers in the worldwide worth chain incorporation. They had the option to get out the importance of ICT in horticultural development yet the exploration is restricted to Sri Lanka henceforth there is a need to prove the results by evaluating what's going on in different nations for steadfastness.

8.4 Harnessing information and communication technology and artificial intelligence in agricultural industry

The mind-boggling connections across subsystems are the place where the test and openings lay. Each of these are in regions where scientists are using large information and AI. For example, an examination meant to expand walnut creation in the southwestern US needs to represent the effect of cultivating rehearses, climate, hereditary qualities and market rehearses as appeared in Fig. 8.5. In particular, a blend of the information learned in every one of the subsystems (HAs, PAs, and SAs) and its application at a more elevated level, where the assorted strings of information can be interlaced for a comprehensive answer is required. Notwithstanding, there is a need to use a comprehensive displaying approach that can represent all factors that sway a cycle and how to survey their interchange (Davenport & Ronanki, 2018).

The acknowledgment of AI in agriculture will require progression on numerous fronts:

- Knowledge extraction: Such strategies will take into account oddity and ordinary example location and expectation of future conditions of the framework. Novel strategies should be produced for improved information combination and extraction of information from heterogeneous sources. These strategies should be strong, effective, and versatile to represent a developing measure of information and to give solid ongoing expectations (Bansal, 2019). For instance, consolidating domesticated animals' areas, wellbeing diagnostics, climate information, and water level detecting to analyze examples of water use.
- Optimized and adaptive decisions: for instance, new methodologies that can decide the condition of a horticultural field, distinguish choices for use as well as progress, do the ideal activity, examine the effect of that activity and refine future choices (Hussain & Al-Turjman, 2020). New methodologies will be expected to recognize the current framework state and conduct to create and apply elective administration plans.

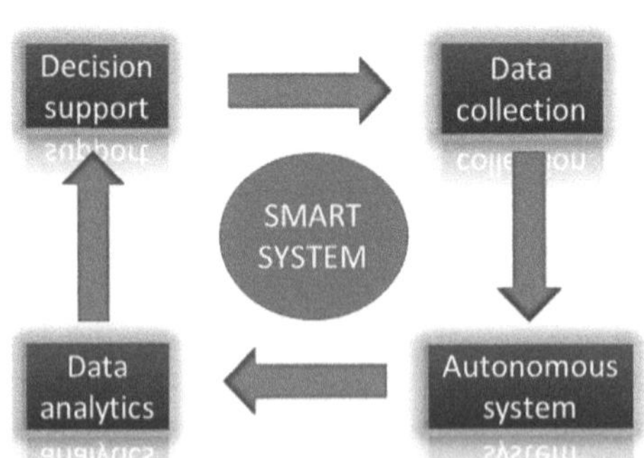

FIGURE 8.5 Illustration of harnessing information and communication technology and artificial intelligence in the agricultural industry.

- Seamless automation and control: if an activity is required, robots particular for explicit assignments can successfully react. Self-ruling robots and other detecting instruments can procure ongoing data that can be utilized, for instance, to screen crop development, recognize infection and bugs, assess water status, and survey creature government assistance. Adequately coupling mechanization with AI includes recognizing the factors to be observed, deciding the sort of sensor(s) to be utilized, distinguishing activities that can adaptively be tended to, and creating mechanical frameworks able to do naturally playing out the activities distinguished (Venkatesh et al., 2003).
- Security, information protection, and provenance: In the entirety of the above fronts, there must be a supported spotlight on security, security, unwavering quality, and flexibility of the frameworks and the AI SoS. Novel innovations in security, for example, blockchain (Vishwanathan, Misra, & Bos, 2019), (Panwar, Misra, & Vishwanathan, 2019) zero-information verifications, configurable security, accessible encryptions, dispersed trust the board, tough appropriated stockpiling can be utilized to improve security and protection as well as improve trust in the information and its accessibility.
- Smart, associated, and secure networks: Stakeholders should be associated safely to represent the ideal degrees of protection (Fountaine, McCarthy, Saleh, 2019). It will be critical to use accessible remote advances, shrewd gadgets, and applications to make a structure that associates HAs in an SoS safely and accommodate a multidirectional stream of data between HAs, SAs, and PAs.

8.5 Artificial intelligence usage of computer networks in agricultural systems

The most recent pattern of computer networks is the HetNet. Thus, the foundation comprises of different remote access advances with different capacities, requirements, and working capacities towards the arising HetNet. The arising portable organizations are turning out to be progressively powerful, heterogeneous, enormous scope, and deliberately perplexing. More explicitly, in the multi-level framework of the most current arrival of LTE-Advanced (HetNet/Small Cells, 2014), pico-cells, macro-cells, femtocells, just as transfer stations and Remote Radio Heads (RRHs), are visualized. Pico-cells and femtocells are with huge prevalence as of late, and are little size, ease, low-power, client uses access points in structures and houses, and offloading information traffic locally (Al-Turjman & Baali, 2019). Full-scale cell base stations utilize committed backhaul, for giving ensured transmission capacity. Transfers are simply MNO-conveyed passageways directing just signals from the full-scale cell to end clients and the other way around in poor zones and dead spots. The related focus full-scale BS is accountable for control and baseband signal preparation, and some radio hardware is moved into RRHs (Ever et al., 2017). RRHs are miniature, high-force, and low-weight units mounted externally on the customary large-scale BS and associated with it for the most part through a fiber optic link for making a conveyed BS. Subtleties of AI and HetNet are examined and depicted in Fig. 8.6 too. Hence, HetNets can diminish the O/CAPEX of MNOs. Conveying different kinds of cells targets offloading the traffic load using various prerequisites on areas, climate qualities, and client elements, for boosting the range re-utilization adequately. In any case, the intricacy of HetNets brings generous plan difficulties and issues, where the mix of AI-based and the system of "Self-Organization Networks" (SON) with the HetNets can be one viable route towards brilliant future versatile organizations in the agricultural industry (Deebak & AL-Turjman, 2021) (Sankayya, Sakthivel, & Gayathri, 2021).

Artificial intelligence strategies incorporate multi-disciplinary procedures from ML, fuzzy neural networks, bio-inspired algorithms, etc., and they have been widely contemplated and applied to streamline PC frameworks and organizations in assorted situations and muddled conditions generally and in the agricultural industries. They have generally lower intricacy empowered by recursive input-based learning and neighborhood cooperation, and henceforth quicker speed of discovering imperfect arrangements contrasted with customary strategies (Zhang, Long, Wang, & Dressler, 2014), (Aliu, Imran, Imran, & Evans, 2013). It has been demonstrated that AI strategies can accomplish exceptional execution (Aliu et al., 2013), as a large portion of them are roused from nature discoveries or spurred by the perspectives of people. Accordingly, the joining of AI procedures into the plan of smart framework becomes one promising pattern for adequately tackling issues in HetNets for the agricultural industry for example, cell arranging,

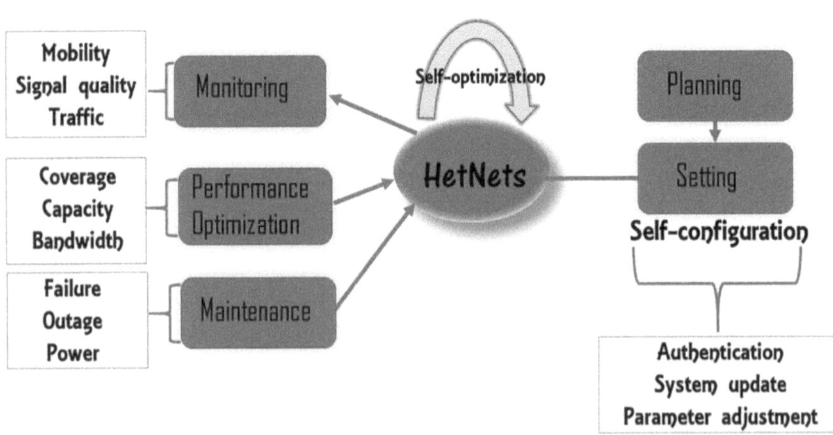

FIGURE 8.6 Illustration of artificial intelligence and HetNet technique in the agricultural industry.

management of resources, adaptation to non-critical failure, and streamlining, and so forth. From the part of decreasing the operational and capital use of MNOs, AI-based procedures can significantly lessen human contribution in operational assignments, and upgrade network limit, inclusion, and Quality of Service (QoS) in HetNets for the agricultural industry, as indicated by the high-level highlights talked about in Self-Organizing Networks (SONs) along following bearings (Imran, Zoha, & Abu-Dayya, 2014; Zhang et al., 2013):

- Self-recuperating, where cells and organizations can naturally identify and recuperate from disappointment and even execute remuneration components at whatever point disappointments happen.
- Self-setup, where recently sent high assortment of cells are automatically configured and refreshed before entering activity undertakings.
- Self-advancement, where cell frameworks can quantify the organization conditions and enhance the settings to improve the presentation as far as inclusion, transmission capacity, obstruction evasion, and QoS, while the comparative issues of framework versatility and energy preservation are also discussed.

8.6 Impacts of information and communication technology and artificial intelligence in the agriculture industry

Simulated intelligence-based frameworks are continually advancing and their groundbreaking impacts have and will keep on having a gigantic effect on society. While a contention can be made concerning the plausibility of AI-related strategies because of the arising idea of the field. Along these lines, besides the specialized issues, it is basic to plan for AI in farming regarding its effect on arrangement, legitimate, and financial territories. It is critical to get ready for the impacts of AI on society as summed up here:

- Exploitation and coordinated effort: How can establishments work together and influence the ability to exploit the inalienable upper hands of every partner. Furthermore, it will be critical to deciding income sources to help not just fundamental AI exploration to propel the field, yet in addition to help the social logical investigation into AI's effects on society (Kahneman, & Tversky, 1979). Creating key organizations to diminish the dependence on the personal responsibility of private elements will be important.
- Regulation: As AI-frameworks are created, guideline changes should be actualized to limit expected negative effects. Another significant territory that should be tended to is the way to manage exclusive innovation, which may obstruct the democratization of AI. For instance, what security, protection, and wellbeing issues should be coordinated into the plan of AI-frameworks, how to manage AI frameworks that can get private data from accessible public information, how to accomplish a harmony

between specialized, legitimate, social, and different territories that shield protection as AI-frameworks are created (Camerer & Weber, 1992).

- Multi-partner sway estimation: Focusing on conventional monetary measurements, for example, Return on Investment to assess the AI-based speculations is as yet significant. Building up a comprehensive system to survey the effect of AI-put together arrangements concerning agribusiness is of high significance. The estimation structure ought to likewise move past the individual farming financial units and assess the effect overall production network framework (Cubitt, Van De Kuilen, & Mukerji, 2018). In any case, supplementing these regular financial exhibition markers with operational ones, for example, following the effectiveness, profitability and quality-related proportions of the horticultural monetary units' pre and post-AI-selection will be significant.
- Employment and schooling: AI will probably affect work dislodging, including low, medium, and conceivably high aptitude occupations. This is especially significant in the farming business, where a test exists to pull in more youthful specialists. Along these lines, it is imperative to recognize approaches to limit such interruptions. For instance, how to manage AI frameworks that may offer delicate types of assistance that, when performed by individuals, may require preparation and affirmation (Etner, Jeleva, & Tallon, 2010; Hussain, Al-Turjman, & Sah, 2021; Hussain, Al-Turjman, Gemikonaklı, & Kirsal Ever, 2021; Hussain, Dawood, & Al-Turjman, 2021; Hussain & Dimililer, 2021). Horticulture is a significant monetary motor and endeavors intending to these interruptions through the improvement of strategy and instructive projects are basic.

8.7 Challenges and discussion

Expert frameworks are devices for farming administration since they can give site-explicit, incorporated, and deciphered advice. Even though AI has made some striking improvements in the agrarian area, it has a beneath-the-normal effect on the horticultural exercises when contrasted with its possibilities and effects in different areas. Notwithstanding, the advancement of master frameworks for agribusiness is genuinely later, and the utilization of these frameworks in business horticulture is uncommon to date (Rajotte et al., 1992). All the more actually should be done to improve horticultural exercises utilizing AI as there are numerous constraints to its usage.

8.7.1 Big data

The strength of an insightful specialist is likewise estimated on the volume of info information. The framework should sift through a significant part of the approaching information. A constant AI framework needs to screen a colossal volume of information. However, it should stay receptive to significant or surprising occasions (Washington & Roth, 1989). The improvement of a horticultural master framework requires the joined endeavors of experts from numerous fields of agribusiness and should be created with the collaboration of the producers who will utilize them. A top to bottom information on the undertaking of the framework is needed from a field master and truth be told, exceptionally applicable information ought to be utilized to improve the framework's speed and exactness.

8.7.2 Accuracy and time

A significant trait of a clever or master framework is its capacity to execute undertakings precisely in a brief timeframe. A framework postpone influences a client's determination of assignment technique. The greater part of the frameworks miss the mark with respect to time or precision, or even both. Procedure determination is estimated to be founded on a cost work consolidating two variables which are the exactness level managed and the exertion needed to synchronize input framework accessibility. Individuals looking to limit exertion and amplify exactness, pick among three systems which are pacing, programmed execution, and checking (Teal & Rudnicky, 1992).

8.7.3 Implementation technique

Since it utilizes large information, the technique for turning upward and preparing ought to be appropriately characterized for speed and precision. The excellence of any master framework lies in its execution system.

8.7.4 Flexibility

Adaptability is a solid property of any stable AI framework. This requires adaptability of the subsystems themselves (Mowforth & Bratko, 1987). It is seen that much advancement has been made in applying AI strategies to specific disengaged errands, yet the significant topic at the main edge of the AI-based mechanical technology innovation is by all accounts the interfacing of the subsystems into a coordinated climate. It ought to likewise have broad capacities to oblige more client information from the field master.

8.7.5 High expense of information

Most AI frameworks are web-based, which thus decreases or confines their utilization, especially in far-off or provincial territories. Likewise, a type of direction will truly assist ranchers with adjusting the utilization of AI on the homestead (Panpatte, 2018). The public authority can uphold ranchers by planning a web administration empowering gadget with lower duty to particularly work with the AI frameworks for ranchers.

8.8 Conclusion

In this paper, we have recognized the possibilities of farming. Through a survey of the advances, we featured different and vast applications of these advances. To this end, AI and ICTs are perceived to be perhaps the most encouraging instruments to help the area. In the review, we have found specialized boundaries that subvert data usage for water system executives. We further examine different zones in which we can apply AI and ICT techniques in the rural business. Likewise, future undertakings can be applied in the expanded help for innovative work endeavors in the use of AI in horticulture. In addition, we have accentuated the difficulties of these advancements. Pilots, enormous scope exploration, and application units will provide information for AI devices, their association with end-clients, and training endeavors. Simultaneous advancement of strategies and standards for protected, secure, and dependable utilization of AI in horticulture. Secure and versatile framework improvements that adopt a comprehensive framework strategy to encourage adaptability and fast reaction to changes in water, atmosphere, and work accessibility are discussed. This overview will provide 90% effectiveness for the understudies and experts studying it.

References

Abawi, G. S., & Widmer, T. L. (2000). Impact of soil health management practices on soilborne pathogens, nematodes and root diseases of vegetable crops. *Applied Soil Ecology*, *15*(1), 37–47.

Abdullah, F. A., & Samah, B. A. (2013). Factors impinging farmers' use of agriculture technology. *Asian Social Science*, *9*(3), 120, Anoop, M., Ajjan, N. and Ashok, K.R., 2015. ICT based market information services in Kerala determinants and barriers of adoption. Economic Affairs, 60(1), p. 117.

Aker, J. C., Ghosh, I., & Burrell, J. (2016). The promise (and pitfalls) of ICT for agriculture initiatives. *Agricultural Economics*, *47*, 35–48, [CrossRef].

Al-Turjman, F., and Baali I. (2019). Machine learning for wearable IoT-based applications: A survey, Wiley transactions on emerging telecommunications technologies. Available from: https://doi.org/10.1002/ett.3635.

Aliu, O. G., Imran, A., Imran, M. A., & Evans, B. (2013). A survey of self organisation in future cellular networks. *IEEE Communications Surveys and Tutorials*, *15*(1), 336–361, First Quarter.

Al-Turjman, F. (2019). The road towards plant phenotyping via WSNs: An overview, Elsevier Computers &. *Electronics in Agriculture*, *161*, 4–13.

Aubry, C., Papy, F., & Capillon, A. (1998). Modelling decision-making processes for annual crop management. *Agricultural Systems*, *56*(1), 45–65.

Awokuse, T. O. 2009 Does agriculture really matter for economic growth in developing countries? In *The American Agricultural Economics Association Annual Meeting*, Milwaukee, Newark, USA.

Badiene, O. (2008). Sustaining and accelerating Africa's agricultural growth recovery in the context of changing global food prices, IFPRI Policy Brief 9.

Balleda, K., Satyanvesh, D., Sampath, N. V. S. S. P., Varma, K. T. N., Baruah, P. K. (2014). Agpest: An efficient rule-based expert system to prevent pest diseases of rice & wheat crops. In *8th International Conference on Intelligent Systems and Control*, Coimbatore, India.

Bansal, G. (2019). *Time to revisit Sumantra Ghoshal's legacy: One AI algorithm at a time*. Taylor & Francis.

Baseca, C. C., Sendra, S., Lloret, J., & Tomas, J. (2019). A smart decision system for digital farming. *Agronomy*, *9*, 216, [CrossRef].

BEA. (2018). Value added by industry as a percentage of gross domestic product [Online]. Available from: https://apps.bea.gov/iTable/iTable.cfm?ReqID = 51&step = 1#reqid = 51& step = 51&isuri = 1&5114 = a&5102 = 5.

Behera, B. S., Das, T. K., Jishnu, K. J., Behera, R. A., Behera, A. C., & Jena, S. (2015). E governance mediated agriculture for sustainable life in India. *Procedia Computer Science*, *48*, 623–629.

Block, S., Timmer, C. (1994). Agriculture and economic growth: Conceptual issues and the kenyan experience. Harvard Institute for International Development.

Camerer, C., & Weber, M. (1992). Recent developments in modeling preferences: Uncertainty and ambiguity. *Journal of Risk and Uncertainty*, *5*, 325–370, [CrossRef].

Cavazza, F., Galioto, F., Raggi, M., & Viaggi, D. (2018). The role of ICT in improving sequential decisions for water management in agriculture. *Water*, *10*, 1141, [CrossRef].

Ceo, W.S.; Foundry, W.; Webb, R. (2019). DigitalWater industry leaders chart the transformation journey [Online]. Available: https://iwa-network.org/wp-content/uploads/2019/06/IWA_2019_Digital_Water_Report.pdf (accessed 09.07.19).

Clara Eli-Chukwu, N. (2019). Applications of artificial intelligence in agriculture: A review [Online]. Available: http://www.etasr.com (accessed 26.08.20).

Cubitt, R., Van De Kuilen, G., & Mukerji, S. (2018). The strength of sensitivity to ambiguity. *Theory and Decision*, *85*, 275–302, [CrossRef].

Davenport, T. H., & Ronanki, R. (2018). Artificial intelligence for the real world. *Harvard Business Review*, *96*(1), 108–116.

Debaeke, P., & Aboudrare, A. (2004). Adaptation of crop management to water limited environments. *European Journal of Agronomy*, *21*(4), 433–446.

Deebak, Dr. B. D., & AL-Turjman, F. (2021). Privacy-preserving in smart contracts using blockchain and artificial intelligence for cyber risk measurements. *Journal of Information Security and Applications*, *58*, 102749. Available from https://doi.org/10.1016/j.jisa.2021.102749.

Dekle, R., & Vandenbroucke, G. (2012). A quantitative analysis of China's structural transformation. *Journal of Economic Dynamics and Control*, *36*(1), 119–135.

Elshorbagy, A., & Parasuraman, K. (2008). On the relevance of using artificial neural networks for estimating soil moisture content. *Journal of Hydrology*, *362*(1–2), 1–18.

Etner, J., Jeleva, M., & Tallon, J. (2010). Decision theory under ambiguity. *Journal of Economic Surveys*, *26*, 234–270, [CrossRef].

Ever, E., et al. (2017). Modelling green HetNets in presence of failures for dynamic large-scale applications: A case-study for fault tolerant femtocells in smart cities. *Elsevier Computer Networks Journal*, *128*, 78–93.

Fan, M., Shen, J., Yuan, L., Jiang, R., Chen, X., Davies, W. J., ... Zhang, F. (2012). Improving crop productivity and resource use efficiency to ensure food security and environmental quality in China. *Journal of Experimental Botany*, *63*(1), 13–24.

Fountaine, T., McCarthy, B., & Saleh, T. (2019). *Building the AI-powered organization. Harvard Business Review*. Harvard Business Publishing.

Furman, C., Roncoli, C., Crane, T. A., & Hoogenboom, G. (2011). Beyond the "fit": Introducing climate forecasts among organic farmers in Georgia (United States). *Climate change*, *109*, 791–799, [CrossRef].

Galioto, F., Chatzinikolaou, P., Raggi, M., & Viaggi, D. (2020). The value of information for the management of water resources in agriculture: Assessing the economic viability of new methods to schedule irrigation. *Agricultural Water Management*, *227*, 105848, [CrossRef].

Galioto, F., Raggi, M., & Viaggi, D. (2017). Assessing the potential economic viability of precision irrigation: A theoretical analysis and pilot empirical evaluation. *Water*, *9*, 990, [CrossRef].

Giupponi, C. (2014). Decision support for mainstreaming climate change adaptation in water resources management. *Water Resources Management*, *28*, 4795–4808, [CrossRef].

Gupta, J. (2019). The role of artificial intelligence in agriculture sector [Online]. Available: https://customerthink.com/the-role-of-artificial-intelligence-in-agriculture-sector/

Haworth, B. T., Biggs, E., Duncan, J., Wales, N., Boruff, B., & Bruce, E. (2018). Geographic information and communication technologies for supporting smallholder agriculture and climate resilience. *Climate*, *6*, 97, [CrossRef].

Henry-Nickie, M., Frimpong, K. and SunFriday, H. (2019). Trends in the information technology sector [Online]. Available: https://www.brookings.edu/research/trends-in-the-information-technology-sector/

Henten, E. J. V., Hemming, J., Tuijl, B. A. J. V., Kornet, J. G., Meuleman, J., Bontsema, J., ... Os, E. A. V. (2002). *An autonomous robot for harvesting cucumbers in greenhouses*. Springer.

HetNet/Small Cells. 2014. [Online]. Available: http://www.3gpp.org/hetnet

Hussain, A.A. & Al-Turjman, F. (2020). Resource allocation in volunteered cloud computing and battling COVID-19. Available from: https://doi.org/10.1201/9781003098881-2.

Hussain, A.A., Al-Turjman, F., Gemikonaklı, E. & Kirsal Ever, Y. (2021). Design of a navigation system for the blind/visually impaired. Available from: https://doi.org/10.1007/978-3-030-69431-9_3.

Hussain, A.A. & Al-Turjman, F. & Sah, M. (2021). Semantic web and business intelligence in big-data and cloud computing era. Available from: https://doi.org/10.1007/978-3-030–66840-2_107.

Hussain, A.A., Bouachir, O. Al-Turjman F. and Aloqaily, M. (2020). AI techniques for COVID-19, in IEEE Access, 8, pp 128776–128795. Available from: https://doi.org/10.1109/ACCESS.2020.3007939.

Hussain, A.A. & Dawood, B. & Al-Turjman, F. (2021). Application of AI techniques for COVID-19 in IoT and big data era: A survey. Available from: https://doi.org/10.1007/978-3-030-60188-1_9.

Hussain, A.A. & Dimililer, K. (2021). Student grade prediction using machine learning in IoT era. 10.1007/978-3-030–69431-9_6.

Imran, A., Zoha, A., & Abu-Dayya, A. (2014). Challenges in 5G: How to empower SON with big data for enabling 5G. *IEEE Network*, *28*(6), 27–33, Nov./Dec.

Irrigants d'Europe Water Framework Directive Position Paper. (2018). pp. 1–9. [Online]. Available from: https://www.irrigantsdeurope.eu/files/documents/IE%20%20WFD%20%20position%20paper%20.pdf (accessed 21.10.20).

Jeuland, M., Hansen, K., Doherty, H., Eastman, L. B., & Tchamkina, M. (2019). The economic impacts of water information systems: A systematic review. *Water Resources Economics*, *26*, 100128, [CrossRef].

Ji, B., Sun, Y., Yang, S., & Wan, J. (2007). Artificial neural networks for rice yield prediction in mountainous regions. *Journal of Agricultural Science*, *145*(3), 249–261.

Johnston, B. F., & Kilby, P. (1975). *Agriculture and structural transformation: Economic strategies in late-developing countries*. Oxford University Press.

Kahneman, D., & Tversky, A. (1979). Prospect theory: An analysis of decision under risk. *Econometrica: Journal of the Econometric Society*, *47*, 263–292, [CrossRef].

Kekane, M. A. (2013). Indian agriculture-status, importance and role in Indian economy. *International Journal of Agriculture and Food Science Technology*, *4*(4), 343–346.

Kimpe, C. R. D., & Morel, J. L. (2000). Urban soil management: A growing concern. *Soil Science*, *165*(1), 31–40.

Kirchhoff, C. J., Lemos, M. C., & Engle, N. L. (2013). What influences climate information use in water management? The role of boundary organizations and governance regimes in Brazil and the U.S. *Environmental Science & Policy*, *26*, 6–18, [CrossRef].

Kolhe, S., Kamal, R., Saini, H. S., & Gupta, G. K. (2011). An intelligent multimedia interface for fuzzy-logic based inference in crops. *Expert Systems with Applications*, *38*(12), 14592–14601.

Kuznets, S. (1973). Modern economic growth: Findings and reflections. *American Economic Association*, *63*(3), 247–258.

Lal, H., Jones, J. W., Peart, R. M., & Shoup, W. D. (1992). FARMSYS-A whole far machinery management decision support system. *Agricultural Systems*, *38*(3), 257–273.

Li, M., & Yost, R. (2000). Management-oriented modelling: Optimizing nitrogen management with artificial intelligence. *Agricultural Systems*, *65*(1), 1–27.

Lopez, E. M., Garcia, M., Schuhmacher, M., & Domingo, J. L. (2008). A fuzzy expert system for soil characterization. *Environment International*, *34*(7), 950–958.

Masseroni, D., Arbat, G., & De Lima, I. P. (2020). Editorial—Managing and planning water resources for irrigation: Smart-irrigation systems for providing sustainable agriculture and maintaining ecosystem services. *Water*, *12*, 263, [CrossRef].

Meza, F. J., Hansen, J. W., & Osgood, D. (2008). Economic value of seasonal climate forecasts for agriculture: Review of ex-ante assessments and recommendations for future research. *Journal of Applied Meteorology and Climatology*, *47*, 1269–1286, [CrossRef].

Molina-Maturano, J., Speelman, S., & De Steur, H. (2020). Constraint-based innovations in agriculture and sustainable development: A scoping review. *Journal of Cleaner Production*, *246*, 119001, [CrossRef].

Moran, M. S., Inoue, Y., & Barnes, E. M. (1997). Opportunities and limitations for image-based remote sensing in precision crop management. *Remote Sensing of Enviroment*, *61*(3), 319–346.

Mowforth, P., & Bratko, I. (1987). *AI and robotics: Flexibility and integration*. Cambridge University Press.

Munirah, M. Y., Rozlini, M., Siti Y. M. (2013). An expert system development: Its application on diagnosing oyster mushroom diseases. In *13th International Conference on Control, Automation and Systems*, Gwangju, South Korea.

Munyua, H., & Adera, E. (2009). Emerging ICTs and their potential in revitalzing small-scale agriculture. *Agricultural Information Worldwide*, *2*(1), 3–9.

National Academies of Sciences Engineering Medicine. (2019). *Science breakthroughs to advance food and agricultural research by 2030*. Washington, DC: The National Academies Press.

Nesheim, I., Barkved, L. J., & Bharti, N. (2017). What is the role of agro-met information services in farmer decision-making? Uptake and decision-making context among farmers within three case study villages in Maharashtra, India. *Agriculture*, *7*, 70, [CrossRef].

Oyakhilomen, O., & Zibah, R. G. (2014). Agricultural production and economic growth in Nigeria: Implication for rural poverty alleviation, Quarterly. *Journal of International Agriculture*, *53*(3), 207–223.

O'Mahony, B., Dalrymple, J., Levin, E., & Greenland, S. (2016). The role of information communications technology in sustainable water management practice. *International Journal of Sustainable Agricultural Management and Informatics*, *2*, 79, [CrossRef].

Pagliai, M., Vignozzi, N., & Pellegrini, S. (2004). Soil structure and the effect of management practices. *Soil and Tillage Research*, *79*(2), 131–143.

Pande, N., & Deshmukh, P. (2015). ICT: A path towards rural empowerment through telecommunication, egovernance, and E-agriculture. *IBMRD's Journal of Management & Research*, *4*(2), 47–54.

Panpatte, D. G. (2018). Artificial intelligence in agriculture: An emerging era of research. Anand Agricultural University.

Panwar, G., S. Misra, and R. Vishwanathan (2019). BlAnC: Blockchain-based anonymous and decentralized credit networks. In *Proceedings of the Ninth ACM Conference on Data and Application Security and Privacy* (pp. 339–350).

Pilarski, T., Happold, M., Pangels, H., Ollis, M., Fitzpatrick, K., & Stentz, A. (2002). *The demeter system for automated harvesting*. Springer.

Rajotte, E.G., Bowser, T., Travis, J.W., Crassweller, R.M., Musser, W., Laughland, D. ... Sachs, C. (1992). Implementation and adoption of an agricultural expert system: The Penn state apple orchard consultant. In *International Symposium on Computer Modelling in Fruit Research and Orchard Management*, ISHS.

Ramli, S. A., Samah, B. A., Hassan, M. S., Omar, S. Z., Bolong, J., & Shaffri, H. A. M. (2015). Potential benefits of ICT for youth agro-based entrepreneurs in Malaysia. *Journal of Applied Sciences*, *15*(3), 411.

Roser, M. (2013). Employment in agriculture [Online]. Available from: https://ourworldindata.org/employment-in-agriculture

Rotz, S., Duncan, E., Small, M., Botschner, J., Dara, R., Mosby, I., ... Fraser, E. D. G. (2019). The politics of digital agricultural technologies: A preliminary review. *Sociologia Ruralis*, *59*, 203–229, [CrossRef].

Sankayya, M., Sakthivel, Rk, Gayathri, N., et al. (2021). Wireless sensor network–based delay minimization framework for IoT applications. *Personal and Ubiquitous Computing*, [Online]. Available. Available from https://doi.org/10.1007/s00779-020-01517-w.

Singh, K.M., Kumar, A., & Singh, R.K.P. (2015). Role of information and communication technologies in Indian agriculture: An overview. Available at SSRN 2570710.

Singh, V., Sankhwar, S., & Pandey, D. (2015). The role of information communication technology (ICT) in agriculture. *Global Journal of Multidisciplinary Studies*, *3*(4).

Snehal, S. S., & Sandeep, S. V. (2014). Agricultural crop yield prediction using artificial neural network approach. *International Journal of Innovative Research in Electrical, Electronics, Instrumentation and Control Engineering*, *2*(1), 683–686.

Syers, J. K. (1997). Managing soil for long-term productivity. The Royal Society.

Syrquin, M. (1988). *Patterns on structural change, Handbook of development economics* (1). Elsevier.

Teal, S.L., Rudnicky, A.I. (1992). A performance model of system delay and user strategy selection. In *Conference on Human Factors in Computing Systems*, California, USA.

Tumbo, S. D., Mwalukasa, N., Fue, K., Mlozi, M. R. S., Haug, R., & Sanga, C. A. (2018). Exploring information seeking behavior of farmers' in information related to climate change adaptation through ICT (CHAI). *International Review of Research in Open and Distributed Learning*, 19, [CrossRef].

Venkatesh, V., et al. (2003). User acceptance of information technology: Toward a unified view. *MIS quarterly*, 425–478.

Verdouw, C. N., Robbemond, R. M., & Wolfert, J. (2015). ERP in agriculture: Lessons learned from the Dutch horticulture. *Computers and Electronics in Agriculture*, *114*, 125–133.

Vishwanathan R., P.G., Misra S., Bos S. (2019). SAMPL: Scalable auditability of monitoring processess using public ledgers. In *ACM Conference on Computer Communication Security (CCS)*.

Vogel, J., Letson, D., & Herrick, C. (2017). A framework for climate services evaluation and its application to the Caribbean Agrometeorological Initiative. *Climate Services*, *6*, 65–76, [CrossRef].

Washington, R., Roth, B. H. (1989). Input data management in real-time AI system. In *11th International Joint Conference on Artificial Intelligence*, Michigan, USA.

Weed Science Society of America. Facts about weeds [Online]. Available from: http://wssa.net/wp-content/uploads/WSSA-Fact-SheetFinal.pdf

World Bank. (2017). ICT in agriculture (Updated Edition): Connecting smallholders to knowledge. Networks and Institutions (Updated Edition) [Online]. Available from: https://openknowledge.worldbank.org/bitstream/handle/10986/27526/9781464810022.pdf?sequence = 2&isAllowed = y (accessed 22.10.20). [CrossRef].

Zhang, Z., Huangfu, W., Long, K., Zhang, X., Liu, X., & Zhong, B. (2013). On the designing principles and optimization approaches of bio-inspired self-organized network: A survey. *Science China Information Sciences*, *56*(7), 1–28.

Zhang, Z., Long, K., Wang, J., & Dressler, F. (2014). On swarm intelligence inspired self-organized networking: Its bionic mechanisms, designing principles and optimization approaches. *IEEE Communications Surveys and Tutorials*, *16*(1), 513–537, First Quarter.

Zhao, Z., Chow, T. L., Rees, H. W., Yang, Q., Xing, Z., & Meng, F. R. (2009). Predict soil texture distributions using an artificial neural network model. *Computers and Electronics in Agriculture*, *65*(1), 36–48.

Chapter 9

Data protection and privacy preservation mechanisms for applications of IoT in smart grids using AI

Barakat A. Dawood[1,2], Fadi Al-Turjman[2], Adedoyin A. Hussain[1,2] and B.D. Deebak[3]

[1]Computer Engineering Department, Near East University, Mersin, Turkey, [2]Department of Artificial Intelligence Engineering, Research Center for AI and IoT, AI and Robotics Institute, Near East University, Mersin, Turkey, [3]School of Computer Science and Engineering, Vellore Institute of Technology, Vellore, India

9.1 Introduction

Energy management is commonly analyzed through a screening assessment before the indications have developed or when an electrical power grid speculates an irregularity. Early identification and representation of grid-level turns out to be significant to save the lives of individuals. Most distance power transmissions pictured in radiographic imaging are favorable and do not grow in a cooperative way. Consumer participation, power quality, system efficiency, pricing, and value-added services are most generally utilized symptomatic devices for the electrical power grid. A utility-sponsored examination of a power grid can fundamentally suspect a line or transmission. The quality metrics of the smart grid, such as balancing authority, adaptive protection, interrupting switch, and correction factors, are newly arising modalities for grid computing and energy management system. The distribution and transmission control center is essential to accommodating the strengths of distance power transmission, safety, and reliability.

The energy management system is the second driving cause for the distribution of electric power substations. It may commonly analyze the growth rate of transmission control centers over the world. In an asymptomatic workup setting (Kimani, Oduol, & Langat, 2019), grid computing examines the key drivers of the smart grid to measure the automation and connectivity

Sustainable Networks in Smart Grid. DOI: https://doi.org/10.1016/B978-0-323-85626-3.00004-1

of various metering components (Pau et al., 2018). Fortuitously, a measuring device assesses the suspicious activities within a self-sufficient system. In standard practice, the conventional power grid characterizes the discoveries as indicated by the supplementary platforms (Wang, Scaglione, & Thomas, 2010). An integrated information provider breaks down the inconspicuous contrasts among cyber attacks and grid failures to characterize the accessibilities such as network, application, service, and platform. An unusual finding detected in a service-oriented architecture commonly requires an indicative workup, which may incorporate various power system elements such as power transformer, transmission lines, substation, and a power transformer.

The normal performance accounts for a power analyzer to measure the rates of electrical transferal such as 86.9% affectability and 88.9% explicitness (Wang, Chen, & Shi, 2011). This prompts substantial intra-spectator and inter-onlooker variability (Krakauer & Cohan, 2017). The synchronous machine can be named as a strange development, where a few cells tend to grow abnormally. An IoT service provider has become one of the main sources of robust IT connectivity and digitization in recent times. Because of an interconnected network, the delivery of electricity from source to on-demand center is substantially more defenseless to meet the requirements of communication architecture. In the year 2019, 42,260 failures were accounted for with complex interconnections such as generation, transmission, and distribution as shown in Fig. 9.1. The cause of power failures in a power grid happens when the energy distribution begins to deliver the transmission lines to consumers. It multiplies the rate of electrical referral through a power analyzer to conduct the suitability of electrical power to other user devices. The power qualities contain transients, swells, and harmonic distortion to reduce the overall consumption of the energy distribution center. As indicated by (Abbasghorbani, Mashhadi, & Damchi, 2014), with a yearly occurrence pace of 144,000 new instances of circuit breaker operation,

FIGURE 9.1 A global trend of the power grid.

a power system allows the electric grid to connect the business through a network of transmission lines in India.

Although DL-based models are progressively used to investigate grid information, the grid computing framework is a subject of controversy (Bitzer & Gebretsadik, 2013; Fayyaz & Nazir, 2012; Huang et al., 2018). Internet of Things (IoT), Machine learning (ML), and its subdiscipline deep learning (DL), have recently acquired astounding outcomes in the smart grid domain (Baek, Vu, Liu, Huang, & Xiang, 2014; Bansal & Singh, 2016; Yigit, Gungor, & Baktir, 2014). In any case, if such suppositions are not right and the outage portrays a heterogenic development, this can prompt a false-positive conclusion (Babun, Aksu, & Uluagac, 2019; Chaichi, Lavoie, Zarrin, Khalifa, & Sie, 2015; Joseph & Jasmin, 2015; Luo et al., 2015; Rusitschka, Eger, & Gerdes, 2010). For an assistive screening and identification, one foresees nonstop straight development of the power outage. The false analysis brings about an outrageous power blackout, expanded expenditure, and further psychological implications. Even master analyzers neglect to know and decipher the specific situation. Additionally, grid failure is an incredibly important component of system examination. Thus, the collaborative grid is essential to help power analysis in the field of grid computing. In an investigation given in (Liu, Mashayekh, Kundur, Zourntos, & Butler-Purry, 2013; Naveen, Ing, Danquah, Sidhu, & Abu-Siada, 2016), system examination helps in decreasing the false negative rate by 77%. In the current situation, the computer-supported investigation is promptly utilized as pattern recognition programming which helps in recognizing the anomalies (malignant or benign) and imprints them for the power analyzer.

The potential attackers and unintentional errors are intrusive to the requirements of the smart grid such as risk assessment, asset management, and access control. The physical and environmental security includes an operational control system, smart metering, and power substations to identify a set of system interoperability. An energy distribution network in grid installation is the earliest phase of the energy distribution network and is treatable. The entire power system can immediately spread to integrate a two-way flow of electricity. Around 1,000,000 threats are determined to examine three high-level securities such as confidentiality, integrity, and availability. It may intersect the power transmission lines to accelerate the development of efficient transmission of grid planners and operators. It may just be 35% of malicious attacks with a late or progressed stage to prepare the lack of awareness and low metering efficiency. In the beginning phase, the rate of survival can be high; the five-year survival rate at this stage is 81% (Habash, Groza, Krewski, & Paoli, 2013). The prognostic assessment utilizing AI methods is conceivable without a piece of earlier information on the loss of transmission and distribution. The most famous grid computing framework is the integrated cybersecurity protection. It was additionally discovered that the grid computing frameworks are very precise in recognizing the standards and guidelines of system interoperability. Because of the developing prevalence of computing models and devices, different examination strategies and algorithms have been planned by analysts worldwide to dissect the common vulnerabilities. An ML-GRID model is involved in the extraction of the ideal component followed by order using the scholarly machine. Essentially, the idea of IoT, machine learning (ML), and artificial intelligence aims to a great extent to impact the field of renewable energies. The component detail is quantified by the shape, intensity, and texture of the suspected area. An overall system for operations and management costs can be summed up as follows (Dash & Dash, 2016; Gu et al., 2016). Table 9.1 gives an outline of the abbreviations used.

Our primary contribution is as follows:

- Describe key auxiliary components required and the assortment of grid computing data.
- Review the literature regarding IoT and ML and their application in grid computing prediction.
- Orchestrate the different methodologies for IoT and ML techniques in diagnosing and predicting power failures.
- Illustrate and discuss the experimental results of grid operations to withstand uncontrolled separation, and cascading failures.

TABLE 9.1 Abbreviations.

Terms	Meaning
ICT	Information and communication technologies
ML	Machine learning
CNN	Convolutional neural networks
CEHD	Canny edge-based edge histogram descriptor
ANN	Artificial neural network
NN	Neural networks
PC	Personal computer
PSF	Point spread function
VS	Virtual studio
MLP	Multilayer perceptron
KNN	k-nearest neighbor
2D	Two-dimensional
SVM	Support vector machine
RAM	Random access memory

9.2 Literature review

A detailed clarification of reviewed research works, algorithms, and models for the analysis of grid resilience is introduced. Histological tests have verifiably been utilized to evaluate heterogeneity, yet both the essential recurrence hazard and the response screen need accurate, non-intrusive estimations. An intelligent network can influence an efficient treatment to ensure a sustainable environment. IoT devices with ML To measure the intra-failure heterogeneity in a grid by applying Spatiotemporal Wavelet Kinetics (SWK) from dynamical comparison-enhanced magnetic reverberation imaging (Kumari, Vekaria, Gupta, & Tanwar, 2020; Siddiqui, Qureshi, Chowdhry, & Uqaili, 2019). The techniques listed above still disapprove of precise analysis of power failure. The pharmaco-active measures are first partitioned into homogeneous sub-districts. Consequently, this paper gives a Deep Learning Framework that combines an algorithm of crucial handling and an autoencoding neural network to arrange various highlights inside the profile of gene articulation effective diagnosis of transmission loss. The high rate of power transmission is essential because of its pervasiveness and fundamentally unique electrical results in intrusive grid computing systems.

Power failure is an important cause of transmission breakdown to affect the complex interconnections. Improving the exactness of the diagnosis is critical and gets one of the vital zones of research. In (Sheikhi, Rayati, Bahrami, Ranjbar, & Sattari, 2015) built a new GPMKL method focused on the utilization of heterogeneous information that incorporates complex interconnections such as quality articulation, adjustment of duplicate numbers, and safeguarding appearance that majorly depends on multiple kernels of learning. Presently, numerous numerical models are proposed for the forecast of grid failures. GPMKL is proposed to execute the component fusion that is joined in the classification of electric transmission lines to highlight regulatory commissions. Supervised learning techniques are utilized to compare their performances to determine which classifier is best for the power grid dataset in (Sha, Alatrash, & Wang, 2016) used a C4.5 Decision tree strategy and was able to achieve 94.74% classification accuracy, then again, (Wang et al., 2018) utilized the RIAC strategy and acquired 96% precision.

The outcome demonstrated SVM-RBF with 96.84% precision when contrasted with another classifier model (Zhan et al., 2014). Various classifiers that have been analyzed so far, which include Naïve Bayes, C4.5, K-nearest neighbor (KNN), and support vector machine (SVM). Utilizing ensemble learning to predict the survival time in breast cancer shows a better precision on the power grid dataset compares to the past outcome. SVM has ended up being the ideal classifier with an accuracy of 96.99% (Tu, He, Shuai, & Jiang, 2017). In (Ma, Zhao, Zhou, & Gao, 2018) utilized ensemble learning and accomplished 97.42% precision. With 97% accuracy, 97.2% review, and 97% accuracy score SVM is the best classifier utilized on the power grid dataset. In (Kotsiopoulos, Sarigiannidis, Ioannidis, & Tzovaras, 2021), they utilized SVM, Random Forest, and Bayesian Network classifiers for grid analysis. In (Kaygusuz, Babun, Aksu, & Uluagac, 2018), they utilized the power-rule technique for computing analysis and they announced 98.2% accuracy utilizing this strategy. In (Bashir et al., 2021) they presented another deep learning way to deal with the short-term risk assessment model. A sum of 270 negative assessment cases was gathered during this research. These cases were partitioned into 200 cases arbitrarily and a 70-case test set. One hundred thirty-five cases were positive and 135 cases stayed negative in the following successive screening of cascading failures.

A grid computing system for profound analysis was then created for the risk assessment comprising of two modules: the versatile module for characterizing the highlights and the danger prediction module. The outcomes indicated that the new failure risk model offered a positive prescient value of 69.2% and a negative prescient estimation of 74.2%, with the general prescient accuracy of 71.4%. A multilayer perception (MLP) classification structure is presented in the forecast of risk to estimate the probability of the power loss getting short-term perceivable grid failures. Authors in (Haque et al., 2020) built up a numerical strategy that promptly measure HER2, a biomarker that characterizes the eligibilities of the transmission lines in smart grid deployment. This research shows that Deep Learning Diagnosis (DLD) can make transmission choices on power quality simpler by recognizing cases with maximum damage to the key component of the power grid. The 12 conflicting cases were then inspected exclusively, and asymptomatic change was made for the initial assessment to improve the connectivity of networks in eight cases. Automated scoring showed 83% agreement with the investigation of power loss ~71% (Babar, Tariq, & Jan, 2020). The diagnostic inconsistency was primarily caused by perceptive contrasts for the evaluation of HER2 expression because of the high changes in HER2 staining.

9.3 Methodology

This is a formal description and it portrays the proposed framework as a whole, incorporated with the objective that it underpins thinking which amalgamates the architecture, conduct and design of the framework which includes parts of the framework and improvements of the framework. This chapter includes the conduct, architecture, and more viewpoints on the system that will work connectedly at the hip while utilizing the overall system, and it is depicted in Fig. 9.2.

9.3.1 System description

The power grid is viewed as one of the most dominant electric sources and a significant cause of failure among the communication infrastructure is due to

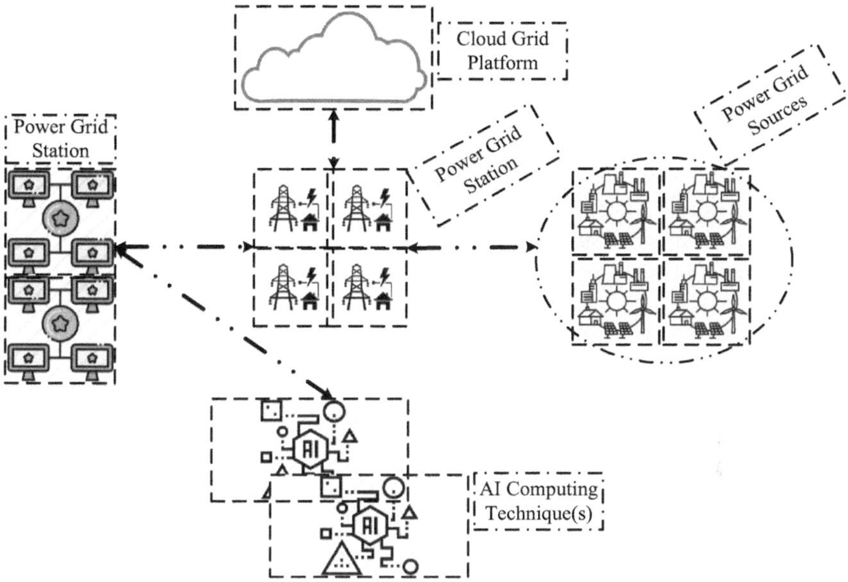

FIGURE 9.2 The architecture of the proposed system.

switching transients. The prediction is additionally significant for power failure underwent by the grid procedure to treat the effects of overload lines or transformers. Even though it is practically difficult to forestall system loss, the early discovery of grid failure during the beginning phases is associated with a reduction in the rate of transmission loss and dismalness of the chain reaction (Cui, Qu, Gao, Xie, & Yu, 2020; Ge et al., 2015; Gholami et al., 2018; Gupta, Kambli, Wagh, & Kazi, 2014; Hofer & Fahringer, 2007; Wu et al., 2011). The proposed framework can be implied to be explicitly created to protect the power flow against the network failure. Normally, the analyzer provides a definitive solution with a prediction of the presumable result of the technical loss, depending on the kind of demand curve, the outcome of the test, the pace of fraud detection development, including memory loss, power supply problem, and data errors. When the inspection is finished, extra tests may also be carried out to figure out the effectiveness of software corruption.

The framework is intended to provide rich, complete inspection and prediction to the system loss about power quality. The recognized kinds of power loss may generally have a magnificent long-haul forecast, particularly if the failure is identified early. However, to achieve a more profound comprehension of power quality instruments, specialized grid deployment is constructed using AI algorithms. It additionally involves the transmission lines which includes the overall information of grid deployment, for example, the study of transmission lines, energy distribution, and power generating sources (Gbadamosi & Nwulu, 2020; Kumar, Saket, Dheer, Holm-Nielsen, & Sanjeevikumar, 2020). As the chapter aims to presume to malign or benign to foresee intermittent or non-repetitiveness of the transmission lines, various standards and guidelines are supported to identify the barriers of smart grid deployment. The designed framework is constructed in two sections, such as AI and IoT sections which are depicted in Figs. 9.3 and 9.4 individually. The IoT section makes use of a smart grid framework where the consumer and the analyzer can have distant conversations keenly without the loss of an electrical power network. The AI section applies AI and deep learning to identify whether high-quality power can be obtained to achieve better energy consumption or not.

The AI section can be subdivided into

- Machine learning (SVM, KNN, and DT) and deep learning (CNN).

  IoT section can be subdivided into

- Smart health system.

9.3.2 Dataset description

The malignant growth is from the power grid dataset. It has instances with the numbers 569 and 32 of features like counting ID and an objective

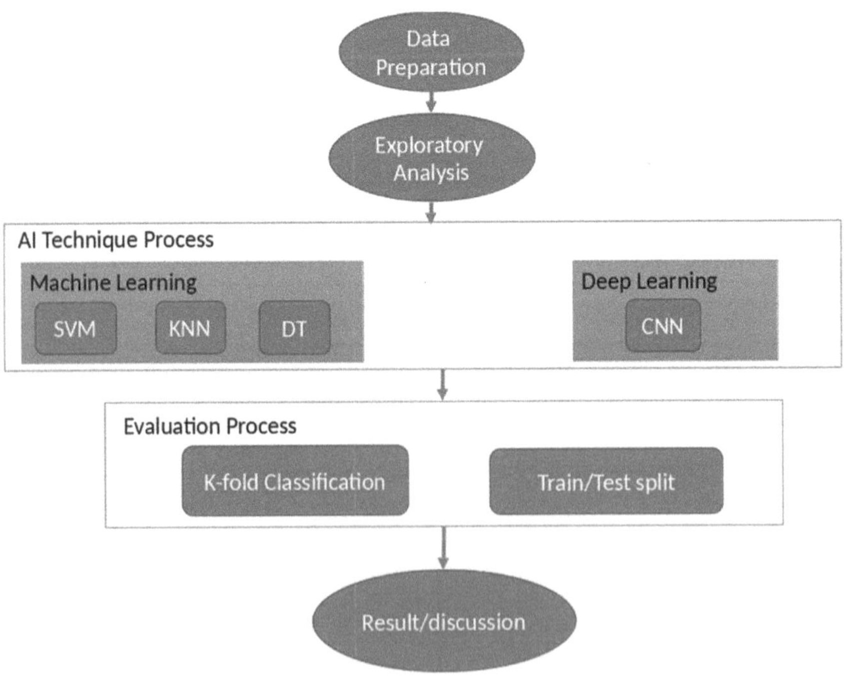

FIGURE 9.3 Description of the proposed AI system.

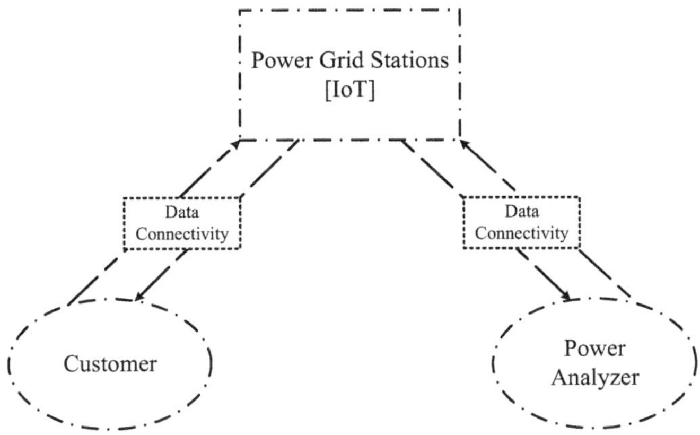

FIGURE 9.4 Description of the proposed IoT system.

variable. The datasets utilized the grid data. Also, the forecast dataset has 4 missing traits to incorporate ⟨10⟩ credits taken from every cell core just as ID and the analysis for $B \rightarrow$benign and $M \rightarrow$malignant. The qualities comprise Id, radius_mean, inspection, texture_mean, area_mean, perimeter_mean, smoothness_mean, concavity_mean compactness_mean, etc. Also, the grid data is extracted in the CSV record (Scott, Wade, & Fernandes, 2020). A genuine arrangement issues such as imbalanced information circulation, for example, spam/extortion location and grid information were handled. The information was kept in a file which establishes a multidimensional and vectors grids binary format. The factors and their portrayal can be found in Table 9.2. The dataset is exceptionally slanted to processes 151 non-repeated results and 47 repeated results. The pre-processing technique are as follows:

- Data pre-handling
  It helps to profoundly comprehend the datasets and conclude how to control the factors that consistently sway the result. Pre-handling and exploratory data analysis are the main concerns for each information venture (Kavitha, 2021).
- Data normalization
  Information normalization is trailed by information examination and perception. For ensuring that information is inside reliable; that is, every information type has a similar substance and organization.

9.3.3 Visualization

The crucial motivation behind representation is portraying the dataset graphically to safeguard the private information of the customers. This is with the innovative psyche to reveal the consumer and prosumer side networks that utilize visual instruments to upgrade and automate the integrity of consumer data

TABLE 9.2 Attributes of the dataset and its meaning.

Attributes	Meaning
Outcome	Whether the power failure has recurred (R) or non-recurred (N)
Id	The Id number of the grid station
Failure node status *(non-existent for inspection dataset)*	Number of positive axillary failure nodes
Power loss size	Diameter of the grid index in centimeters
Texture	Standard deviation of the gray-scale values
Area	Area of the power leakage
Symmetry	Value of symmetry
Radius	Distance from center to points on the perimeter
Perimeter	Perimeter of the grid area
Concave points	Number of concave portions of the contour
Fractal dimension	Coastline approximation—1, measure of how complicated the power loss is
Concavity	Severity of concave portions of the contour
Smoothness	Local variation in radius lengths, how round the quality area is
Compactness	(Perimeter squared divided by area)-1

including billing and pricing. The pattern of information perception is portrayed as stacking information into the application, information depiction, and structure confirmation, showing the outcome, a pattern of depiction, assembling, and investigating the information to refine the privacy requirements of smart grid networks. The yield factors extract the possible outcome segments to exploit the learning mechanisms with error connectivity over network rings.

9.3.4 Computational environment

VScode is a no-pay and a standard programming condition comprising of a strong set-up of instruments for data assessment and real strategies. The investigations utilized the VScode, which is an open-source condition to drive the use of SL procedures. Python is one of the most well-known programming tools and offers numerous libraries to deal with information science Tensorflow for VScode is a library to characterize numerous exhaustive capacities. It uses experienced language among the gathering individuals to examine the utilization factor. Specifically, the library utilizes (DT), (KNN),

(SVM), and (CNN) packages to aggregate the generation of energies recorded by the prosumers. Pandas, SciPy, Keras, NumPy, TensorFlow, Matplotlib, Scikit-Learn, were used as the libraries to relate a large number of fault tolerances. The activity was assessed on a personal computer with 2.3 GHz, RAM: 8GB.

9.3.5 Techniques used

Support vector machine (SVM): It is a supervised learning algorithm that utilizes classification and regression issues. They are particularly on high dimensional spaces and productive since they utilize a subset of support vectors. They are a bunch of supervised learning techniques utilized for classification, regression, and outliers' discovery. It uses a precision rate to forecast the corrupted meters of the huge dataset. It comprises hypothetical and numeric capacities to highlight the regression issues. AI strategy depends on 3D and 2D to demonstrate an incentive-based demand-response system. SVM gathers basic examples from all categories to derive a linear function that isolates the energy utilization data. SVM is one of the supervised classification models applied in the region of grid computing.

KNN: K-nearest neighbor algorithms characterize the information by discovering its nearest neighbors in a multidimensional component space occupied by known models from a preparation informational collection. The outcomes of this algorithm could differ depending upon two techniques such as Manhattan distance and Euclidian distance to analyze the incentive claims.

DT: Decision tree is one of the most broadly utilized classification techniques. A decision tree comprises hubs that have precisely one approaching edge, aside from the root hub. It has no approaching edges to exploit the resistant features of the data aggregation scheme. Decision trees are exceptionally easy to comprehend, decipher and improve human meaningfulness. The blend of parameters is applied on 10-crease cross-validation to examine the precision of integer factorization 93.62% with a particularity of 90.66% on the discovery dataset. A hub with active edges is an interior hub, while other hubs are called leaves or terminal hubs, or decision hubs. Irregular decision trees can decipher and straightforwardly deal with insignificant traits.

CNN: Convolution neural networks (CNNs) are separated from the single shrouded layer neural network as far as their profundity. The further deep into the network, the more unpredictable the highlights the hubs can identify and learn. In profound learning networks, every layer trains on an unmistakable arrangement of highlights dependent on the preceding layer's result.

Smart power grid station: IoT is associating the actual world with the virtual universe of the Internet. The actual world incorporates the primitive features of the power grid such as geographical information, marketing system, customer services, running state of equipment, power quality, and load monitoring. There are possible utilizations of grid services IoT to analyze efficiency ratio. Applying IoT innovations to grid services helps to improve the nature of an individual's life, the degree of a constant outage, fault cautioning, and line breakage.

9.3.6 Experimental procedure

To analyze the energy demands effectively, the information preparing strategies, playing out the novel AI techniques are critical. Since it is an AI venture, ML and deep learning measures are utilized in numerous grid enterprises. The means underneath show a rundown of the ML and deep learning technique. By then, it endeavored with different evaluations and attributes to apply the execution of the AI technique trailed by the gathering strategy to conclude which model can fit best using the confusion matrix and heat map. The IoT steps are also portrayed to discuss the significance of AI implementation. Table 9.3 shows the analytical steps involved in power grid stability and load balancing using learning mechanisms.

9.4 Results

To assess the viability of the learning techniques, the dataset was tested with various AI models. A bunch of different investigations was done to arrange the hyper-parameters' classifier. Each model and feature assume a basic function to obtain the higher assessment model. The models are integrated with various settings to accomplish the most noteworthy accuracy. Three ML models such as SVM, KNN, and DT were preferred to predict the information with various arrangements of training tests and one DL model in particular CNN. Similarly, an enormous informational collection of 569 instances and 32 traits are used in this examination. The informational index is part into two sections the training and testing sets. Each model has the yield to analyze the load stabilities and power grid failures. When the best foreseeing model was distinguished, the preparation set was diminished in size to perceive a preferable limit to anticipate the best outcome. Then, the best-distinguished classifiers and assessment metrics were expressed toward the beginning of the task. VScode was used to explore different functional libraries to analyze the grid parameters. Each model used a similar area of informational collections. As follows, the result of the models is elucidated in the procedure section. Table 9.4 depicts a segment of the input of the power grid dataset.

9.4.1 Experimental parameters

These are the used parameters to show how valid the techniques are in predicting grid failures and load instabilities.

TABLE 9.3 Analytical steps involved in power grid stability and load balancing using learning mechanisms.

Step-1	Choosing datasets and distinguishing the undertaking issue.
Step-2	Cleaning up the datasets, recoding factors, scaling factors, and dealing with missing factors were implemented. Examine the power grid dataset to pick, stack and arrange.
Step-3	Plots of circulation and different plots were assumed that can help to determine prediction of grid failures. The exploratory examination was also performed to decide whether there are any exceptions or anything of interest to affect the transmission cycle.
Step-4	Next, the preparation cycle was initialized to the grid stabilities.
Step-5	The testing predominantly records the load balancing to contrast the outcomes and to check if the outcomes are substantial, train/test split of 80% to 20% test to is utilized to examine the split proportionate
Step-6	From the k-fold characterization, it was found that we can apply another strategy for joining models. Next assessment measures were applied to the writing audit techniques, for example, confusion matrix and heat map.
Step-7	In the learning strategies, the datasets were part into 80% to 20% train and test, and afterward, various blends were smoothed out, upgraded and results were yield.
Step-8	Results were assembled and a similar investigation was finished. At the last unique assessment, measures were determined and enhancement of boundaries of the AI strategy was performed.
The steps of the grid IoT can be summarized as follows:	
	Start the process
Step-1	Initialize the application interface through a dedicated wired/wireless port
Step-2	Navigate the power grid failures to locate the geographic locations
Step-3	The remote station can verify the grid information and record the load stability and transmission failure
Step-4	End the process

The F1 score conveys the balance between the precision and recall.

$$F_1 = 2 \times \frac{\text{precision} \times \text{recall}}{\text{precision} + \text{recall}} = \frac{\text{TP}}{\text{TP} + \frac{1}{2}(\text{FP} + \text{FN})} \tag{9.1}$$

where TP is the number of true positives, FP is the number of false positives, and FN is the number of false negatives.

TABLE 9.4 Breast cancer data depiction.

Digit	Mean texture	Mean perimeter	Mean area	Mean smoothness
0	17.99	10.38	122.80	1001.0
1	20.57	17.77	132.90	1326.0
2	19.69	21.25	130.00	1203.0
3	11.42	20.38	77.58	386.1
4	20.29	14.34	135.10	1297.0

Precision—Precision calculates how precise/accurate our model is out of those predicted positives, and how many of them are positive. Precision is a good measure to determine when the costs of False Positive are high.

$$\frac{\text{True Positive}}{\text{True Positive} + \text{False positive}} \tag{9.2}$$

Recall—Recall calculates how many of the Actual Positives our model captures through labeling that are Positive (True Positive).

$$\frac{\text{True Positive}}{\text{True Positive} + \text{False Negative}} \tag{9.3}$$

Support—Support is the number of actual occurrences of the class in the specified dataset. Imbalanced support in the training data may indicate structural weaknesses in the reported scores of the classifier and could indicate the need for stratified sampling or rebalancing. Support does not change between models but does inspect the evaluation process.

Accuracy—This approach calculates the accuracy of our classification model.

$$\frac{\text{True Positive} + \text{True Negative}}{\text{True Positive} + \text{True Negative} + \text{False positive} + \text{False Negative}} \tag{9.4}$$

Weighted average—The precision and recall for each class, and weight by the number of instances of each class. That will give you the *weighted* precision and recall.

$$\text{weighted}_p = \frac{(\text{Pc1}^*|\text{c1}|) + (\text{Pc2}^*|\text{c2}|)}{|\text{c1}| + |\text{c2}|} \tag{9.5}$$

Macro average—It takes the average of the precision and recall of the system on different sets.

$$\text{Macro average precision} = \frac{P1 + P2}{2} \tag{9.6}$$

$$\text{Macro average recall} = \frac{R1 + R2}{2} \tag{9.7}$$

9.4.2 Results discussion and comparison

The outcomes of the ML models are illustrated in the form of figures and tables. To guarantee model consistency, all models utilized a train/test splitting of 80% to 20% proportion. The table comprises the output of usage of the models on the diagnosis of power grid failures and load instabilities. From the evaluation, the best model for grid computing diagnosis was the (SVM) (support vector machine) acquiring an exactness of 97% whereas KNN as 96% and DT 93%. Additionally, the models were played out with default boundaries to accomplish the SVM as the model with the best accuracy followed by KNN and DT (Al-Turjman, Ulusar, & Nawaz, 2020). The distinction in pre-handling and train/test split was productive for each model to exhibit accuracy with default boundaries from the start.

In the classifier, the genuine class was marked with 1 and the bogus class was named with 0. The sections of the confusion matrix were summed up by the class forecasts. Matrix rows sum up the genuine qualities inside each class. True Positive is a positive class perception (1), which is effectively named positive by the model (Hussain & Al-Turjman, 2021). True Negative is a negative class perception that is accurately named negative. False Positive is a negative class perception (0), which is mistakenly delegated positive. False Negative is a positive class perception that is mistakenly delegated negative. However, for the KNN, the best value for K was intentionally left with no specific value (Hussain, Dawood, & Al-Turjman, 2021). Afterward, the best value of K had a deduction for us to enhance the accuracy and efficiency of the proposed module. Fig. 9.5 shows the best value of k in the KNN model. It can deduce after the mapping that the best value of K was 8. Thus, it is proven that the model can be improvised to enhance the accuracy of the deployed model (Hussain, Al-Turjman, & Sah, 2021). In Fig. 9.6, it is observed that the change in the confusion matrix may give an efficient result to inspect the load stabilities. Table 9.4 shows the experimental results for improved KNN using the normal distribution after taking the best value for the nearest neighbors to increase the accuracy rate from 94% to 96% efficiently.

After implementing the same method as above, it is witnessed that the SVM model can perform better. This shows durability of this model for diagnosis. Fig. 9.7 shows the heat map and confusion matrix of the SVM model and it is noticed that a change in the predictability of the model is accurate and efficient to analyze the load stability and power grid failures. Table 9.5 shows the experimental result of the F1 score, recall, and precision

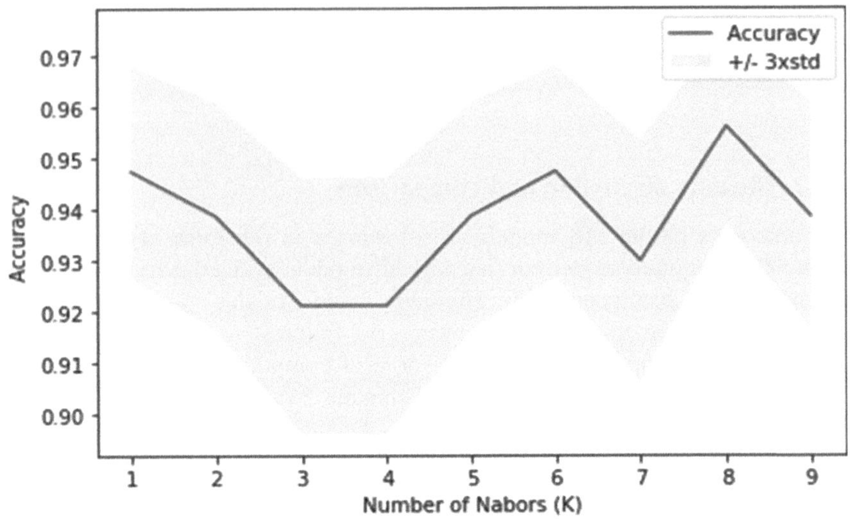

FIGURE 9.5 Depicting the best value of K in the K-nearest neighbor model.

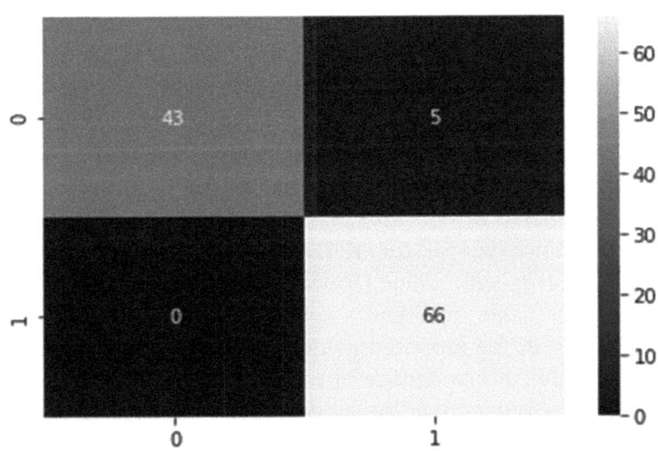

FIGURE 9.6 Depicting the heatmap and confusion matrix of the improved K-nearest neighbor model.

(Hussain, Dawood, & Al-Turjman, 2021). It is also observed that it had an accuracy of 96% without further improvisation. The comparative result with the previous model shows the same even after improvising the best k value of the KNN. Fig. 9.7 shows the improved result of the SVM model, which can also show a change in the confusion matrix deducing the enhanced model better than before. Table 9.5 shows the improvisation and normalization of the SVM classifier that infers the accuracy set of the SVM model to

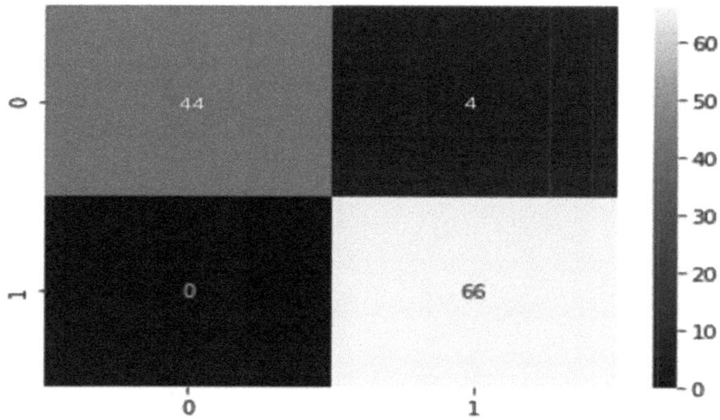

FIGURE 9.7 Depicting the heatmap and confusion matrix of the improved support vector machine model.

TABLE 9.5 Experimental result of the improved K-nearest neighbor.

	F1-score	Recall	Support	Precision
1	0.96	1.00	66	0.93
0	0.95	0.90	48	1.00
Weighted average	0.96	0.96	144	0.96
Macro average	0.95	0.95	144	0.96
Accuracy	0.96		144	

increase from 96% to 97% by the optimization factors such as Gamma and C Parameters.

Table 9.6 shows that the SVM model is the best classifying model. It gives the highest accuracy 97% than the KNN with 96% accuracy scores respectively. The result mentioned was further classified after the enhancement of the models by normalization. It is also compared with the next ML model known as the DT as shown in Fig. 9.8. The figure shows the tree diagram of the predictability to improve the effectiveness of the previous models. The DT proved not to be effective after hitting an accuracy of 93%. Moreover, it is still better compared to the deep learning model Fig. 9.9.

Deep learning in different ML includes object identification, and grouping to demonstrate the advancement innovation of AI and IoT. The best organizations accomplish barely higher proficiency and surpass all different errands as far as precision when estimated utilizing a similar informational index. Detail learning in strategies from the info data concerning the

TABLE 9.6 Experimental result of the improved support vector machine model.

	F1-score	Recall	Support	Precision
1	0.97	1.00	66	0.94
0	0.96	0.92	48	1.00
Weighted average	0.96	0.96	144	0.97
Macro average	0.96	0.96	144	0.97
Accuracy	0.97		144	

FIGURE 9.8 Depicting the tree of the DT model.

objective yield following traditional Machine Learning techniques includes a handmade element extraction level. The presentation of the framework is subjected to the nature of the division given, which complies with time and exertion. The design strategies offer a detailed contribution to separated the load instabilities with power grid failures. Fig. 9.10 shows the exactness of

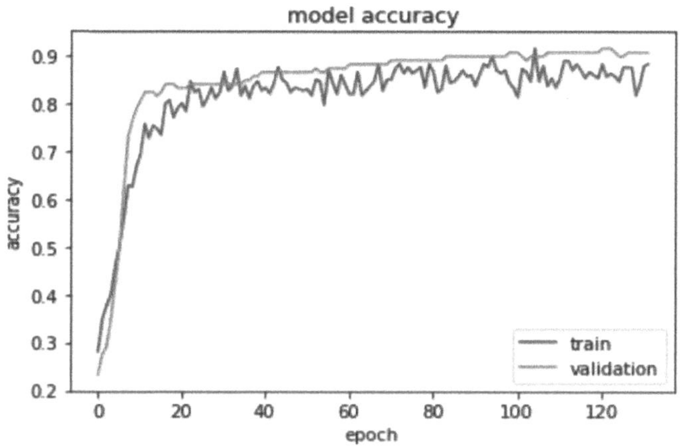

FIGURE 9.9 Illustration of graphical model accuracy of the CNN.

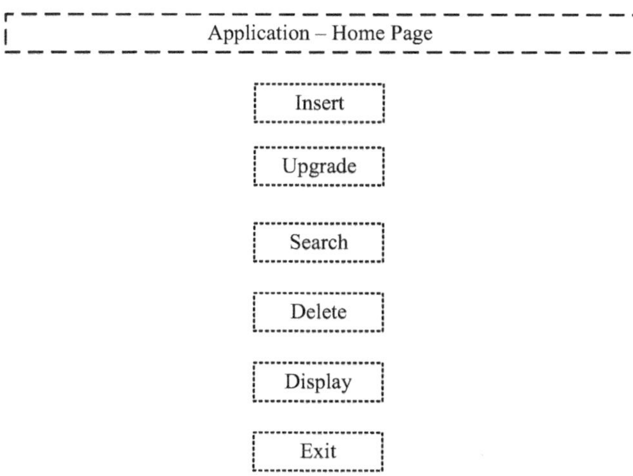

FIGURE 9.10 Illustration of the smart grid analyzer.

the proposed CNN. Fig. 9.9 shows the model accuracy of the CNN. Meanwhile, the absence of any CNN results brings about an improved start to finish grid grouping framework. However, the accuracy and loss are compared in the images described above. The above perception shows that the strategy being proposed is less helpless to disproportionate informational collections and is without a doubt steadier, contrasted with other categories of ML which show dramatic changes in the different exactness levels. Anyway, Table 9.7 shows poor accuracy of 89% compared to the other Ml models, but almost similar to that of the DT. The accuracy image shows a spike in

TABLE 9.7 Experimental result of the CNN model.

	F1-score	Recall	Support	Precision
1	0.92	0.73	66	0.86
0	0.84	0.99	48	0.98
Weighted average	0.89	0.89	144	0.91
Macro average	0.88	0.86	144	0.92
Accuracy	0.89		144	

the accuracy as the number of epochs increases, and with minimal loss as the epochs increase.

For transmission loss and load instability with grid computing, far-off observation utilizing the IoT smart wellbeing application is highly recommended. With a smart grid, practitioners could distinguish fluctuation or deviations from gauge measures and alarm the warnings or remote inspection to check in or look for quick recovery. This could provide adequate intensification management to generalize wellbeing and long haul results after malignant growth treatment, as the IoT system extends the possible ways to measure, assess, and screen the power grid.

In general, it is worth noting that more accuracy does not fit the grid proposals and treatment choices to malignancy of the growth of hazards. IoT information might have the option to improve our comprehension of the microenvironment's commitment to examine the predictive features. Fig. 9.10 shows the landing page of the smart wellbeing application where the grid analyzer can explore immediately from the solace of their home or anyplace. This guarantees information sharing between the control center and the grid station. Figs. 9.11 and 9.12 depict the data list and info data of a grid structure. To genuinely incorporate emergency facilities with grid computing, there is a need to use innovative IoT applications. It may expand the interest of power administrations around the world to upgrade connections, generation, and distribution of power sources.

Three diverse AI models, such as KNN, SVM, and DT and one profound learning model like CNN were applied to analyze a huge dataset, that is, 569 instances and 32 attributes. The models had a short time to show the presence of exactness on power grid data for better diagnosis and prediction. To guarantee model consistency, all models executed a train/test splitting of 80% to 20% proportion. From the analysis, the ML model known as SVM (support vector machine) acquires an exactness of 97% where the exactness of KNN, DT, and CNN are as follows: 96%, 93%, and 89%, respectively. Precision correlation for the AI model for the diagnosis of power grid failure and transmission loss is shown in Table 9.8.

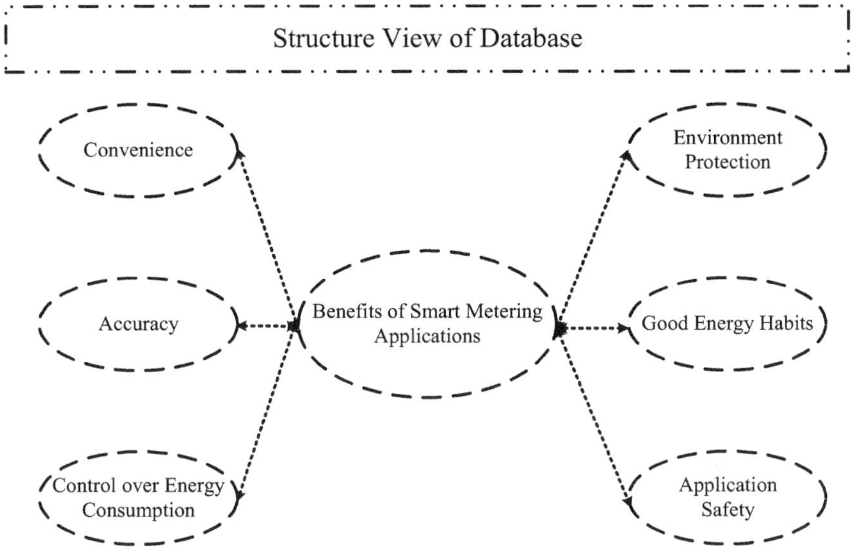

FIGURE 9.11 Illustration of the smart health app information index.

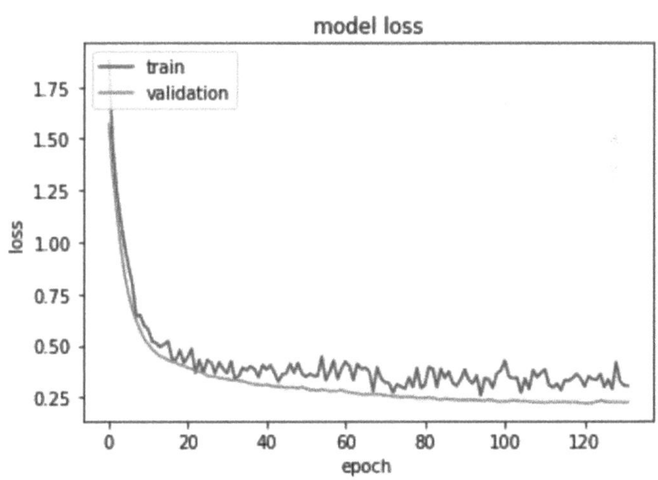

FIGURE 9.12 Illustration of graphical model loss of the CNN.

The created order structure for SVM is demonstrated to be fit for grouping failure or loss proficiently. The pre-owned AI models in this postulation showed promising exhibitions as portrayed in this part. SVM can handle the train/test split to achieve the best outcome compared to other learning models The deep learning model utilizes the CNN precisely in the preparation stage to sum up unperceiving information to inspect the transmission loss as

TABLE 9.8 Experimental result of the AI models.

	SVM	KNN	DT	CNN
Initial accuracy	0.96	0.94	–	–
Accuracy	0.97	0.96	0.93	0.89

the number of epochs increases. Note that the difference in the accuracy was high between both CNN and SVM. Thus, with the outcome deduced from this experiment, it is shown that the control center can utilize both AI and IoT efficiently in predicting and diagnosing the grid failure whenever utilized. This is a wake-up call to all grid organizations to utilize the proposed method to address the distribution/transmission loss.

9.5 Conclusion

Artificial intelligence and IoT have indicated impressive improvement in the diagnosis of power grid failures. The power grid failures proffer a remarkable setting for grid diagnosis by additionally thinking about the operational condition and their reaction to power erection. In any case, there are difficulties for perfect diagnosis and checking of distribution/transmission loss regardless of advanced technologies. This chapter introduces a smart diagnosis and prediction module for power grid failures. It can incorporate a social thread of data to improve the prescient modules. SVM, KNN, DT, and CNN were chosen to analyze the AI modules which use IoT modules to inspect the power loss. These network models were compared with various parameters set to find the accuracy of each model to discover a robust network to classify the best outcome. Usage of unique power grid datasets and a comparison of the outcomes of the techniques utilized the state-of-the-art, convolutional neural networks to solve classification problems. CNNs are additionally known to supplant conventional classification, which includes connected layers to train the handmade components. It may assist in the improvement of the performance of the handmade components.

Perceptions from this trial result demonstrated that it is not advisable to utilize only one AI procedure on datasets with fewer features since an additional decrease of the component vectors may result in misclassification and performance imprecision. The implemented SVM method depicted DL and other ML classifying models to obtain a performance of 97% and the least being the CNN with 89%. Dependent on the measures, the future study will be carried out with different classifications to decide the least subset of distinguished grid dataset. The framework designed presently gives exact diagnoses and prediction, executing it with a neuro-fuzzy control algorithm can also be strongly suggested to utilize different data to assess and boost the efficiency of the recommended algorithm. The discoveries may start up innovative strategies to assess the performance int terms of voting components.

References

Abbasghorbani, M., Mashhadi, H. R., & Damchi, Y. (2014). Reliability-centred maintenance for circuit breakers in transmission networks. *IET Generation, Transmission & Distribution*, *8*(9), 1583–1590.

Al-Turjman, F., Ulusar, U., & Nawaz, M. (2020). Intelligence in the internet of medical things era: A systematic review of current and future trends. *Elsevier Computer Communications Journal*, *150*(15), 644–660.

Babar, M., Tariq, M. U., & Jan, M. A. (2020). Secure and resilient demand side management engine using machine learning for IoT-enabled smart grid. *Sustainable Cities and Society*, *62*, 102370.

Babun, L., Aksu, H., & Uluagac, A. S. (2019). A system-level behavioral detection framework for compromised CPS devices: Smart-grid case. *ACM Transactions on Cyber-Physical Systems*, *4*(2), 1–28.

Baek, J., Vu, Q. H., Liu, J. K., Huang, X., & Xiang, Y. (2014). A secure cloud computing based framework for big data information management of smart grid. *IEEE Transactions on Cloud Computing*, *3*(2), 233–244.

Bansal, P., & Singh, A. (2016, December). Smart metering in smart grid framework: A review. In *2016 Fourth International Conference on Parallel, Distributed and Grid Computing (PDGC)* (pp. 174–176). IEEE.

Bashir, A. K., Khan, S., Prabadevi, B., Deepa, N., Alnumay, W. S., Gadekallu, T. R., ... Maddikunta, P. K. R. (2021). Comparative analysis of machine learning algorithms for prediction of smart grid stability. *International Transactions on Electrical Energy Systems*, e12706.

Bitzer, B., & Gebretsadik, E. S. (2013, September). Cloud computing framework for smart grid applications. In *2013 48th International Universities' Power Engineering Conference (UPEC)* (pp. 1–5). IEEE.

Chaichi, N., Lavoie, J., Zarrin, S., Khalifa, R., & Sie, F. (2015, August). A comprehensive assessment of cloud computing for smart grid applications: A multi-perspectives framework. In *2015 Portland International Conference on Management of Engineering and Technology (PICMET)* (pp. 2541–2547). IEEE.

Cui, L., Qu, Y., Gao, L., Xie, G., & Yu, S. (2020). Detecting false data attacks using machine learning techniques in smart grid: A survey. *Journal of Network and Computer Applications*, 102808.

Dash, S., & Dash, D. K. (2016, November). Serverless cloud computing framework for smart grid architecture. In *2016 IEEE 7th Power India International Conference (PIICON)* (pp. 1–6). IEEE.

Fayyaz, S., & Nazir, M. M. (2012). Handling security issues for smart grid applications using cloud computing framework. *Journal of Emerging Trends in Computing and Information Sciences*, *3*(2), 285–287.

Gbadamosi, S. L., & Nwulu, N. I. (2020). Reliability assessment of composite generation and transmission expansion planning incorporating renewable energy sources. *Journal of Renewable and Sustainable Energy*, *12*(2), 026301.

Ge, Y., Flueck, A. J., Kim, D. K., Ahn, J. B., Lee, J. D., & Kwon, D. Y. (2015). Power system real-time event detection and associated data archival reduction based on synchrophasors. *IEEE Transactions on Smart Grid*, *6*(4), 2088–2097.

Gholami, A., Shekari, T., Amirioun, M. H., Aminifar, F., Amini, M. H., & Sargolzaei, A. (2018). Toward a consensus on the definition and taxonomy of power system resilience. *IEEE Access*, *6*, 32035–32053.

Gu, Y., Jiang, H., Zhang, Y., Zhang, J. J., Gao, T., & Muljadi, E. (2016, September). Knowledge discovery for smart grid operation, control, and situation awareness—A big data visualization platform. In *2016 North American Power Symposium (NAPS)* (pp. 1–6). IEEE.

Gupta, S., Kambli, R., Wagh, S., & Kazi, F. (2014). Support-vector-machine-based proactive cascade prediction in smart grid using probabilistic framework. *IEEE Transactions on Industrial Electronics*, *62*(4), 2478–2486.

Habash, R. W., Groza, V., Krewski, D., & Paoli, G. (2013, August). A risk assessment framework for the smart grid. In *2013 IEEE Electrical Power & Energy Conference* (pp. 1–6). IEEE.

Haque, N. I., Shahriar, M. H., Dastgir, M. G., Debnath, A., Parvez, I., Sarwat, A., ... Rahman, M. A. (2020). Machine Learning in generation, detection, and mitigation of cyber-attacks in smart grid: A survey. *arXiv preprint arXiv*, *2010*, 00661.

Hofer, J., & Fahringer, T. (2007, September). Grid application fault diagnosis using wrapper services and machine learning. In *International Conference on Service-oriented Computing* (pp. 233–244). Springer, Berlin, Heidelberg.

Huang, Y., Lu, Y., Wang, F., Fan, X., Liu, J., & Leung, V. C. (2018, October). An edge computing framework for real-time monitoring in smart grid. In *2018 IEEE International Conference on Industrial Internet (ICII)* (pp. 99–108). IEEE.

Hussain, A. A., & Al-Turjman, F. (2021). Artificial intelligence and blockchain: A review. *Transactions on Emerging Telecommunications Technologies*. Available from https://doi.org/10.1002/ett.4268.

Hussain, A. A., Al-Turjman, F. & Sah, M. (2021). Semantic web and business intelligence in big-data and cloud computing era. https://doi.org/10.1007/978-3-030–66840-2_107.

Hussain, A. A., Dawood, B. & Al-Turjman, F. (2021). Application of AI techniques for COVID-19 in IoT and big data era: A survey. https://doi.org/10.1007/978-3-030–60188-1_9.

Hussain, A. A., Dawood, B. & Al-Turjman, F. (2021). IoT and AI for COVID-19 in scalable smart cities. https://doi.org/10.1007/978-3-030–76063-2_1.

Joseph, S., & Jasmin, E. A. (2015, December). Stream computing framework for outage detection in smart grid. In *2015 International Conference on Power, Instrumentation, Control and Computing (PICC)* (pp. 1–5). IEEE.

Kavitha, R. (2021). *Application of knowledge and data mining to build intelligent systems. Nanoelectronics, circuits and communication systems* (pp. 33–44). Singapore: Springer.

Kaygusuz, C., Babun, L., Aksu, H., & Uluagac, A. S. (2018, May). Detection of compromised smart grid devices with machine learning and convolution techniques. In *2018 IEEE International Conference on Communications (ICC)* (pp. 1–6). IEEE.

Kimani, K., Oduol, V., & Langat, K. (2019). Cyber security challenges for IoT-based smart grid networks. *International Journal of Critical Infrastructure Protection*, *25*, 36–49.

Kotsiopoulos, T., Sarigiannidis, P., Ioannidis, D., & Tzovaras, D. (2021). Machine learning and deep learning in smart manufacturing: The smart grid paradigm. *Computer Science Review*, *40*, 100341.

Krakauer, N. Y., & Cohan, D. S. (2017). Interannual variability and seasonal predictability of wind and solar resources. *Resources*, *6*(3), 29.

Kumar, S., Saket, R. K., Dheer, D. K., Holm-Nielsen, J. B., & Sanjeevikumar, P. (2020). Reliability enhancement of electrical power system including impacts of renewable energy sources: A comprehensive review. *IET Generation, Transmission & Distribution*, *14*(10), 1799–1815.

Kumari, A., Vekaria, D., Gupta, R., & Tanwar, S. (2020, June). Redills: Deep learning-based secure data analytic framework for smart grid systems. In *2020 IEEE International Conference on Communications Workshops (ICC Workshops)* (pp. 1–6). IEEE.

Liu, S., Mashayekh, S., Kundur, D., Zourntos, T., & Butler-Purry, K. (2013). A framework for modeling cyber-physical switching attacks in smart grid. *IEEE Transactions on Emerging Topics in Computing*, *1*(2), 273–285.

Luo, F., Dong, Z.Y., Zhao, J., Zhang, X., Kong, W., & Chen, Y. (2015, July). Enabling the big data analysis in the smart grid. In *2015 IEEE Power & Energy Society General Meeting* (pp. 1–5). IEEE.

Ma, Y., Zhao, F., Zhou, X., & Gao, Z. (2018, August). Summary of cloud computing technology in smart grid. In *2018 IEEE International Conference on Mechatronics and Automation (ICMA)* (pp. 253–258). IEEE.

Naveen, P., Ing, W. K., Danquah, M. K., Sidhu, A. S., & Abu-Siada, A. (2016, March). Cloud computing for energy management in smart grid-an application survey. In *IOP Conference Series: Materials Science and Engineering* (Vol. 121, No. 1, p. 012010). IOP Publishing.

Pau, M., Patti, E., Barbierato, L., Estebsari, A., Pons, E., Ponci, F., ... Monti, A. (2018). A cloud-based smart metering infrastructure for distribution grid services and automation. *Sustainable Energy, Grids and Networks*, *15*, 14–25.

Rusitschka, S., Eger, K., & Gerdes, C. (2010, October). Smart grid data cloud: A model for utilizing cloud computing in the smart grid domain. In *2010 First IEEE International Conference on Smart Grid Communications* (pp. 483–488). IEEE.

Scott, J., Wade, N. S., & Fernandes, J. (2020, October). The need for tidiness in energy data analysis. In *2020 IEEE PES Innovative Smart Grid Technologies Europe (ISGT-Europe)* (pp. 434–438). IEEE.

Sha, K., Alatrash, N., & Wang, Z. (2016). A secure and efficient framework to read isolated smart grid devices. *IEEE Transactions on Smart Grid*, *8*(6), 2519–2531.

Sheikhi, A., Rayati, M., Bahrami, S., Ranjbar, A. M., & Sattari, S. (2015). A cloud computing framework on demand side management game in smart energy hubs. *International Journal of Electrical Power & Energy Systems*, *64*, 1007–1016.

Siddiqui, I. F., Qureshi, N. M. F., Chowdhry, B. S., & Uqaili, M. A. (2019). Edge-node-aware adaptive data processing framework for smart grid. *Wireless Personal Communications*, *106*(1), 179–189.

Tu, C., He, X., Shuai, Z., & Jiang, F. (2017). Big data issues in smart grid—A review. *Renewable and Sustainable Energy Reviews*, *79*, 1099–1107.

Wang, C., Zhu, Y., Shi, W., Chang, V., Vijayakumar, P., Liu, B., ... Fan, Y. (2018). A dependable time series analytic framework for cyber-physical systems of IoT-based smart grid. *ACM Transactions on Cyber-Physical Systems*, *3*(1), 1–18.

Wang, S., Chen, H., & Shi, W. (2011). SPAN: A software power analyzer for multicore computer systems. *Sustainable Computing: Informatics and Systems*, *1*(1), 23–34.

Wang, Z., Scaglione, A., & Thomas, R. J. (2010, December). Electrical centrality measures for electric power grid vulnerability analysis. In *49th IEEE conference on decision and control (CDC)* (pp. 5792–5797). IEEE.

Wu, L. L., Kaiser, G. E., Rudin, C., Waltz, D. L., Anderson, R. N., Boulanger, A. G., ... Pooleery, M. (2011). Evaluating machine learning for improving power grid reliability, https://doi.org/10.7916/D8SB4F1Z.

Yigit, M., Gungor, V. C., & Baktir, S. (2014). Cloud computing for smart grid applications. *Computer Networks*, *70*, 312–329.

Zhan, J., Huang, J., Niu, L., Peng, X., Deng, D., & Cheng, S. (2014, December). Study of the key technologies of electric power big data and its application prospects in smart grid. In *2014 IEEE PES Asia-Pacific Power and Energy Engineering Conference (APPEEC)* (pp. 1–4). IEEE.

Chapter 10

An efficient lightweight privacy-preserving scheme for secondary users in cognitive radio networks

B.D. Deebak[1], Fadi Al-Turjman[2] and Hadi Zahmatkesh[3]
[1]School of Computer Science and Engineering, Vellore Institute of Technology, Vellore, India, [2]Department of Artificial Intelligence Engineering, Research Center for AI and IoT, AI and Robotics Institute, Near East University, Mersin, Turkey, [3]Department of Mechanical, Electronic and Chemical Engineering, OsloMet—Oslo Metropolitan University, Oslo, Norway

10.1 Introduction

An approach known as a database-driven cognitive radio network plays a crucial role to exploring the limited spectrum resources in a grid of everything. This database-driven approach exploits dynamic spectrum sharing to access the privilege of the licensed network channels. It uses secondary users (SUs) exclusively to gain channel access without causing any external interference to the primary users (Jain & Kaur, 2020). Cognitive radio networks (CRNs) share the power spectrum among the primary and secondary users to improve the utilization efficiency of the power spectrum that may alleviate the crisis of spectrum sharing. The potential applications of CRN include vehicular networks, smart cities, and the Internet of Things (IoT) to operate dynamic spectrum sharing in a large-scale IoT system. The large-scale IoT system accesses the alternate power spectrum to solve key issues such as device heterogeneity, interference management, and energy efficiency (Chaudhari, Zennaro, & Borkar, 2020).

A reliable CRN has two basic approaches of secondary users to secure the availabilities of channel information (Tashman & Hamouda, 2020). The former approach is known as sensing technology which cooperates with secondary users to identify the idle channels and subsequently publish the status of the channel to the fusion center. The center will decide whether an idle channel can be in prior to use to avoid interference with the primary users.

Sustainable Networks in Smart Grid. DOI: https://doi.org/10.1016/B978-0-323-85626-3.00011-9

The latter approach is declared as a database-driven approach to manage the roles of commercial entities including Google Inc., and Microsoft. They use spectrum-available information to store the locations of secondary users in the system database (Aswathy, Gopakumar, & Ahamed, 2020). Moreover, this approach executes a database query to gain system access which can collect the availabilities of channel information and transmission parameters to the secondary users. An independent agency of the federal communications commission specifies a primary approach of database-driven CRN to administrate the geolocation database (Khalfi, Hamdaoui, Guizani, & Elmaghbub, 2021).

In a database-driven approach, the CRN is emerging as a key technology to address the issue of spectrum utilization. The cognitive radio-based smart grid network refers to the unlicensed user as the secondary user to gain the privileges of a licensed frequency band (Gupta & Kumar, 2020). The primary user learns spectrum availability to explore the data structures of the spectrum database which probabilistically executes the queries of the secondary users to manage the representation of the system database. The execution of the spectrum database guarantees location privacy to analyze the characteristics of cost performance (Hajian, Ghahfarokhi, Vasfi, & Ladani, 2021). To achieve unconditional security with minimum communication overhead, the proposed LEPP relies on a two-factor authentication scheme. The authentication scheme exploits the compact version of the data contents to gain an understanding of spectrum availability. To learn the query data structures, the secondary user explores an additional architecture that can significantly minimize the computation and communication overhead (Shao & Zhang, 2020).

Internet of Everything (IoE) has emerged with the convergence of IoT technologies to provide pervasive connectivity and network services. The development of information society increases the demands of the social environment to fulfill the constraints of connected IoT devices. In order to meet the business constraints, IoT is undergoing a series of architectural changes. It has no vertical and integrated structure to connect the system, applications, and terminals, whereas the horizontal structure integrates the IoT systems to make the system devices interacting with different network domains. In the grid network, the smart advanced metering system converges as an intelligent system to connect the smart meters and network utilities through advanced metering infrastructure (AMI) (Albayati, Abdullah, Abu-Samah, Mutlag, & Nordin, 2020). It can introduce an enhanced AMI to operate an innovative strategy to prevent cyberattacks and integrate the flow of renewable resources to monitor the connectivity of smart devices without additional system overhead. The storage of electricity rate, power consumption, and an outage can provide a better diagnosis to enable self-care analysis (Bohloul, 2020).

Lately, telematics and the smart grid have transformed conventional diagnosis systems into IoT-enabled smart grid environments to address issues such as consumer privacy and security among interconnected devices. Various sensors and handheld devices that is, smart-bands, smart-device, and tablets may perform roles as monitors, sensors, or communicators to distribute the renewable energies over an electrical grid (Sikder, 2020). The Cisco report reveals that the usage of IoE is expected to generate a net profit value of $14Trillion which highlights costs and revenue of the private sector from 2013 to 2022 (Perelmuter, 2021). Also, this report shows the involved risk factors ~$9.5Trillion to understanding the challenges of industry-specific case models such as smart infrastructure and power grid computing. The computing paradigm integrates with the electrical/power grid and information technologies to distribute the power/energies between the supplier and consumer efficiently. Most infrastructures consider information and communication technology (ICT) to collect and process various types of energy resources through the knowledge of smart meters.

10.1.1 Smart grid: information acquisition system

The smart grid has its own intellectualization of traditional power generation/management systems to integrate high-speed communication networks. It uses advanced sensing and progressive devices, control strategies, decision-support system, and measurement technology to offer an environment-friendly ecosystem (Vinodh, Antony, Agrawal, & Douglas, 2020). At present, smart grid computing is challenging to offer security and privacy to achieve better service digitalization. Moreover, several inefficiencies such as energy, reliability, and power delivery focus on the infrastructure of the existing system, especially in the information acquisition system. The data monitoring system collects the details of the device configuration, however, the system is not relying on the process of device authentication to analyze the malicious behaviors. The smart grid acquisition system restates the communication several times between the collector and the smart meter to record the activities of electronics devices (Ropp, Reno, Bower, Reilly, & Venkata, 2020).

The smart meters actively manage the sensitive information of an electric grid such as voltage, power consumption, and outage to gain access to the business interface. It may upload the sensitive data to the smart meter to issue the control order or to adjust the state of the electric power transmission. The communication interface addresses the issue of malicious transmission to analyze the grid instability and false-controlled instruction. Thus the smart grid analyzes the issues of existing device authentication to enhance the efficiencies of lightweight privacy-preserving mechanisms. The entities, such as smart meters and collectors, may even prefer data integrity and anonymity to offer better credibility among the communication parties. Fig. 10.1 shows a grid architecture of the information acquisition system. The smart grid rapidly discovers a new scenario of smart energies to offer effective access control to realize the precondition of device authentication.

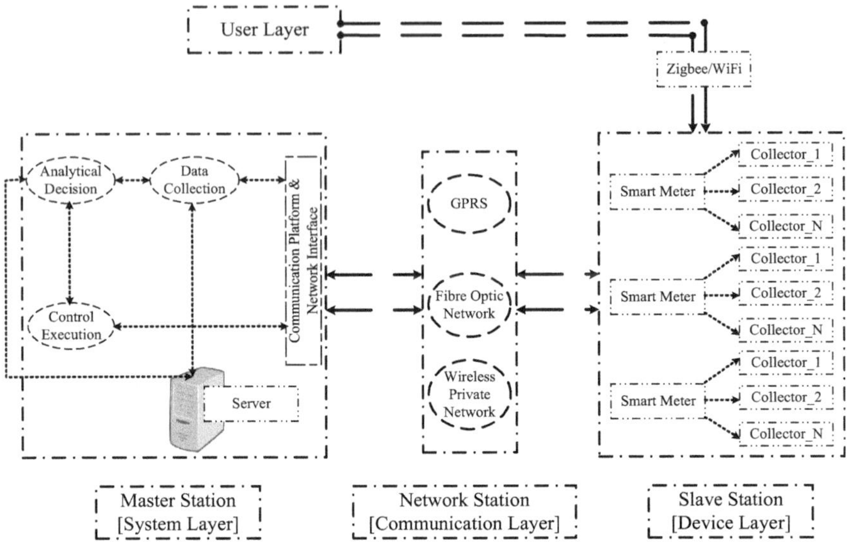

FIGURE 10.1 A grid architecture of the information acquisition system.

10.1.2 Smart grid: smart electrical distribution network/smart breaker

A smart electrical distribution network will diagnose the fault locations to observe the growth of new computing paradigms including smart breaker and control centers (Feng, Wang, Chen, Strbac, & Kang, 2020). The observation of electrical networks may utilize grid data using a wide-area measurement system or supervisory control and data acquisition to monitor the status of any time-synchronized sensor in real-time. As a result, the smart breaker isolates the faulty lines till the network returns to the state of reconfiguration. In order to connect the switches remotely, the smart breaker records the status of the faulty lines to the control center. The control center utilizes demand-side management to balance the supply of electricity or to control the excessive output of the adjacent station. It may automate the process of the smart breaker to implement the design strategies of the smart grid. The control center includes application software to perform faulty analysis which may generate a successive action using an automatic breaker to report the action-based movement to the smart breaker.

It uses a wireless infrastructure to send the action-based movements to the smart grid which may even experience various cyberattacks to maintain system efficiencies (Rahouti, Xiong, & Xin, 2020). Therefore, the smart grid demands a secure path or robust authentication to improve long-distance power transmission and system safety. In real-time, the smart breaker uses ABB – eVD4 to explore the minimum requirements of hardware systems such as Pentium 4, Processor Base-2.4 GHz, RAM-512 MB, and Disk Space-

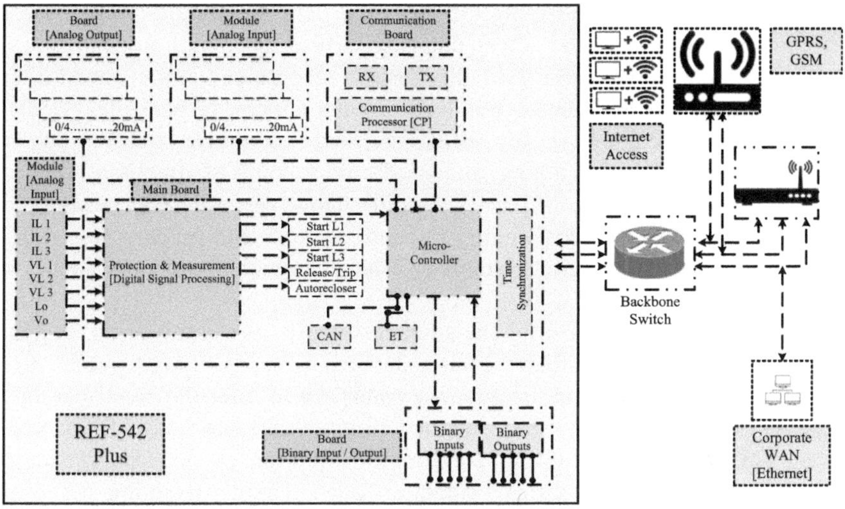

FIGURE 10.2 *REF* 542*Plus*—an automation control unit via a web interface.

40 MB during the initialization phase. The software requirements include Window XP, .Net Framework, Connectivity Package-eVD4 RBX615, and Production Build-PCM600 to classify the eVD4 as intelligent electronic devices. The international standard uses the communication model of IEC-61850 to exploit the controller of the smart breaker, that is, with Modbus and GOOSE Messaging. Fig. 10.2 shows the abilities of *REF* 542 *Plus* (Lowczowski, Lorenc, Zawodniak, & Dombek, 2020) to automate the control unit through a web interface.

An automation control unit uses a web interface to exploit the principle features of IPSec including encryption and authentication. It has a process of key management to administer and manage the cryptography keys in order to handle the abilities of cryptosystems including generation, exchange, and storage of public/private keys. The authentication mechanism assures the legitimacy of the packet transmission using a strategy of encryption algorithm to secure the exchange of confidential data.

10.1.3 Smart grid: emerging technologies

Usually, wearable sensing devices are embedded in the power grid to sense, collect, and store the electrical data to the remote server. Upon the storage of electrical data, system diagnosis and emergency decisions can be accordingly obtained. In data and collaborative grid systems, a power system fault diagnosis is provided with various sensing components to examine the vital signs of the circuit breaker (Pan et al., 2020). In order to exercise realistically, the sensors actively collect and transmits electrical data wirelessly over insecure public networks. After receiving the electrical data, the remote server stores the grid information in the server database. Upon successful authentication, the system analyst examines the faulty data to recommend proper monitoring and preventive measurements (Bakdi, Bounoua, Guichi, & Mekhilef, 2021). As a consequence, emergency situations such as power restore, power restore, inrush current, lightning, and short-circuit could be timely inferred to save the conditional state of the PV module. However, wireless sensing and communication are still challenging to cause security and privacy-preserving issues that is for smart grid systems (Keshk, Turnbull, Sitnikova, Vatsalan, & Moustafa, 2021).

The application scenario of the smart grid networks involves the system entities to connect the smart sensing devices wirelessly over public networks. Generally speaking, an adversary may try to infer the electrical data passively that traces the user's location, login time, and usage frequency. Using a Google search, the above user information can be associated to guess the consumer identities (Mandal, 2020). Upon extracting the user information, the adversary may gain authorized access to exploit features of the communication such as system location, pricing data, utility back office, etc. This situation may even worsen the conditional state of the grid network, where the adversary intercepts the recovery follow-ups to wrongly recommend periodic maintenance. As a result, the smart meter may imprecisely evaluate the performance of channel state information such as leakage, bit error, and bit mismatch rate. On the other hand, the smart meter may experience some security risks such as insider threat, phishing, identity theft, and information extortion to adopt the development of IoT (Kim, Oh, Ryu, & Lee, 2020).

The grid network is openly available to gain user access, thus the feeder automation in the distributed network may be prone to several security issues such as spoofing, eavesdropping, and tampering. Moreover, it may restrict the consumption of resources including energy, storage, and bandwidth to cause high communication and computation costs. Thus a security mechanism prefers advanced IoT technologies to secure the communication between the control center and the distribution network. To analyze security efficiencies, an intelligent computing system is highly preferred. It can integrate some critical components to examine the wired transmission lines, whereby the networks may infer the risk rates in terms of interception, modification, and fabrication. Since user privacy is so essential to protecting device access, the online learning mechanism is chosen to analyze the security features of the networks including public and private (Nguyen et al., 2021). The technological advancement explores various real-time objects to sense and collects a massive amount of sensitive data. Smart grid technologies utilize different communication sources to impart the learning intelligence to classify the specific use of the network applications (Cioffi, Travaglioni, Piscitelli, Petrillo, & De Felice, 2020).

10.1.4 Application of IoT in smart grids

Most of the IoT environment classifies the real-time objects into: (1) Physical devices such as smart home, drone, vehicle, camera, and television; (2) Virtual devices such as e-commerce, e-ticket, and e-wallet to secure the wireless channel; (3) Mitigation of the network overhead including computation and communication to soothe branch power loss; (4) Generation of an anonymous identity among smart breaker and control centers to prevent fabricated forgeries; and (5) Resisting different types of cyberattacks to offer better integrity and device interactivity with hashing function. The major achievement of IoT applications is to provide a better decision-making process without human intrusion. Importantly, it can easily be available on any web-based application system to address challenging issues such as data confidentiality, accessibility, and privacy leakage. In the United Nations conference on climate change (COP21), the smart-IoT plays a crucial role to address the potential hazard of CO_2 emissions (Alsamhi et al., 2021). It is evident that the integration of smart-IoT may bring several benefits to the cities including data management, device navigation, fault diagnosis, load monitoring, and operational maintenance. It uses a principal actor known as a grid station to represent the connectivity as a smarter technology to connect the physical objects extensively to design a grid computing system, namely mobile workforce management, smart metering systems, and digital automation.

Since IoT components have limited resource constraints, most real-time application systems prefer to use cybersecurity approaches (Mohamed, Al-Jaroodi, & Jawhar, 2020). It applies complex cryptography mechanisms to prevent security vulnerabilities. The major challenge is to provide a preventive solution to secure user identities, physical objects/things, and network devices over a public network. The IoT design and development critically measure the vulnerabilities to analyze the impact of trustable networking services. Thus this paper presents a lightweight efficient privacy-preserving (LEPP) scheme to resolve the issues of dynamic time synchronization. Moreover, the proposed LEPP achieves the key features such as availability, integrity, confidentiality, and integrity to authenticate the services of IoT devices in cognitive radio networks.

10.2 Research background

Of late, various research efforts, such as performance modeling, resource optimization, and protocol design, have been initiated to address the issue of location privacy. Despite its significance, privacy issue in cognitive radio networks focuses on the collaborative context of spectrum sensing to signify the scope of database-driven approach. In this approach, a dynamic spectrum auction protects the privacy of secondary users to learn the spectrum opportunities. Most of the grid applications adopt the learning techniques such as differential privacy, k-anonymity, and private information retrieval to emulate a predefined set of probabilistic data structures. Zhang, Fang, Li, Zhu, and Dong (2015) relied on the concept of database query using secondary users to locate the regional information. This approach constructs an energy trade-off between location privacy and service utility to achieve better privacy levels and to minimize the rate of service utilization. Gao, Zhu, Liu, Li, and Cao (2013) designed a private information retrieval to obtain the information from a learning database. Trostle et al. Trostle and Parrish (2010) exploited the pattern of spectrum utilization to identify the potential attacks and to localize the information of a secondary user. Zhang, Zhang, He, and Cheng (2015) applied differential privacy to rely on the mechanism of ϵ-geo indistinguishability to collect the location of the secondary user.

10.2.1 Privacy preserving in cooperative spectrum sensing

Li et al. (2012) discovered the privacy leakage of secondary users to infer the sensing report of location privacy. This mechanism observed the received signal strength to discover the attack of differential location privacy which uses sensing reports to deduce private activities of the secondary users. Shi, Chan, Rieffel, Chow, and Song (2011) presented a privacy-preserving mechanism to utilize secret sharing and privacy aggregation. Their mechanism generates a dummy inspection report to resist a single-report location privacy attack. However, this mechanism could not support fault tolerance to perform fusion analysis. Mao, Chen, Zhang, Wang, and Zhong (2015) developed two schemes using cooperative sensing spectrum such as ElGamal and threshold cryptosystem to prevent a single point of failure and internal malfunctions. However, their scheme consumes more computation and communication costs to initialize a process of non-interactive threshold cryptosystem to decrypt the ciphertext. Grissa, Yavuz, and Hamdaoui (2015) combined Yaos' Millionaires and order-preserving mechanism to generate the encrypted sensing report. The report uses an order-preserving mechanism to learn the encrypted values of received signal strength. Moreover, their scheme employed a network gateway to explore the functionalities of cryptographic tools to realize the privacy-preservation of secondary users.

10.2.2 Cybersecurity in smart grid

In the past, IoT cybersecurity approaches have been initiated for layer-level analysis. It analyzes the cyberattacks to provide proper countermeasures to the entire IoT environment. Cirani, Picone, Gonizzi, Veltri, and Ferrari (2015) designed an external authentication mechanism using OAuth logic that provides an authentication layer to examine network services such as Hypertext Transfer Protocol (HTTP) and Constrained Application Protocol

(CoAP). Hummen, Shafagh, Raza, Voig, and Wehrle (2014) designed a delegated architecture to offload the expensiveness of the datagram transport layer security (DTLS) connection. It has a delegated server to examine the use of certificate-based handshake protocol. Unfortunately, their scheme cannot reduce the transmission overhead to improve the network performance. Moosavi et al. (2015) presented a secure authentication mechanism to analyze the key features of smart e-Health. Aman, Chua, and Sikdar (2017) designed an efficient authentication framework, which uses an unclonable function to examine the security weaknesses using a challenge-response mechanism. Moreover, it allows us to develop various industrial applications that are omnipresent to provide a wide range of service interactions and cooperation. A reliable transmission shows whether the transmission rate is frequently changing on the network connection or not. Significantly, it may address the performance issue of the transmission rate to infer the chances of a double-spending attack (Nicolas & Wang, 2019).

In addition, the consensus is a method, which has been agreed upon by the participating nodes in the blockchain by which new transactions are added to the blockchain. It incorporates the control functions to improve system efficiency, medical standards, productivity, and scientific workflows. Of late, the IoT has gained the researcher's attention for smart application development including smart TV, Internet boxes, automation systems, etc. IoT is subjected to collect the relevant information, that is, from sensor objects, distributed sensor networks, and wearable sensors (Muzammal, Talat, Sodhro, & Pirbhulal, 2020). Generally, the sensing devices exchange real-time services and environment information over distributed networks to build a smart environment (Singhal, Gankotiya, Agarwal, & Verma, 2012). On the other hand, a distributed ledger or database uses a technique of blockchain technology to preserve and record the information of connective objects (Dalby et al., 2018). Blockchain technology provides equal provisional access and authority to connective objects regardless of a master-slave concept.

10.2.3 Smart grid with smart breaker

Grid computing systems allow legal participants to read and write the smart object activities, and in addition, each legal participant has the privilege to check or verify whether the storage information is secure to record. Moreover, the system allows a controller group of participants to read and write the information of connective objects where the participant has a right to validate the data control access (Morris et al., 2011). IoT network is rapidly growing to enhance the range of network connective access that could deal with millions or billions of smart objects to cooperate with each other (Agarkhed & Ashalatha, 2017). Fig. 10.3 shows a branching scenario of a power grid with the exchange control of a smart breaker.

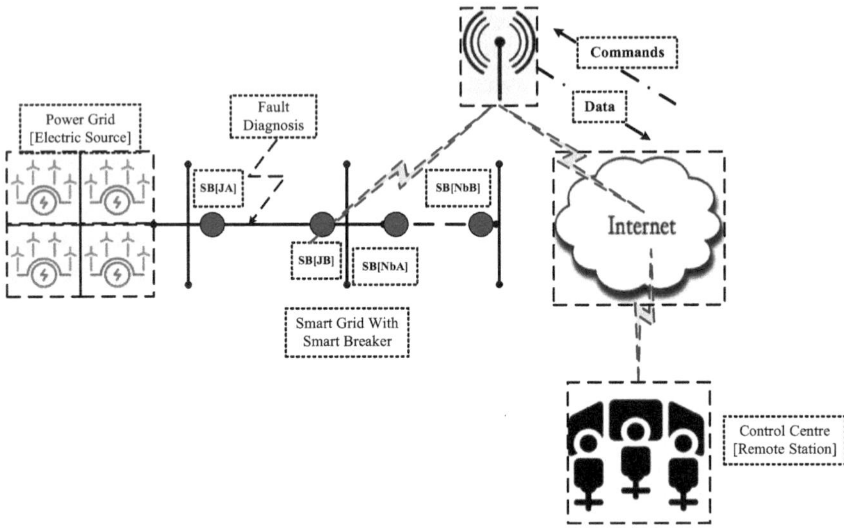

FIGURE 10.3 A branching scenario of a power grid with the exchange control of smart breaker.

Of late, it has gained research attention for the growth of 100 billion smart objects interconnected by 2020 (Tewari & Gupta, 2020). Normally, IoT is based on the heterogeneous nature that might increase the complexity of the security mechanisms (Padmavathi & Shanmugapriya, 2009). Most IoT applications centralize the cloud-based data management system that could experience several harmful attacks namely cyberattacks (Rid & Buchanan, 2015), denial-of-service (Lau, Rubin, Smith, & Trajkovic, 2000), and malware (Wu, Mizuno, & Havinga, 2002). Most importantly, a tiny embedded device cannot integrate or install a security-related middleware to resist the vulnerabilities as it has restricted computation and storage space on the cloud (Jokhio, Ashraf, Lafond, & Lilius, 2013). However, cloud data access provides access privilege between different areas or different purposes to facilitate data analytics. As a result, it may cause severe security risks to ensure data confidentiality, integrity, and access privacy. Hence, a robust lightweight privacy-preserving mechanism is highly recommended to guarantee a secure device interaction between the smart objects. It is evident that security and privacy are the key issues to improve the salient features of the smart grid IoT environment such as energy plan, usage, and status of electricity.

To resist the potential vulnerabilities, the authentication and key agreement mechanism can be integrated. It may ensure better security and privacy through the use of the access control mechanism. In IoT, the integral authentication tools strengthen the specific security features to build a powerful distributed network. These tools devolve the network functionalities to accommodate the changes of autonomous devices.

10.3 Preliminaries

This section discusses the system model and security goals to show the typical environment of smart grid location-based services to meet the standard constraints of smart grid in IoT systems.

10.3.1 Location-based privacy-preserving system

Fig. 10.4 shows the system architecture of a location-based privacy-preserving system in which a location-based privacy-preserving system (Jiang et al., 2021) acquires either a smart device or a fusion center to fix/place the service request to the service provider. The smart device is used to sense and generate the sensing report of the power grid. The important condition parameters are as follows: customer information, meter data, transformer load, load forecasting, and resource planning. These parameters are deliberately utilized in the adoption of advanced smart metering to measure the key terms such as energy demand, consumption, and current usage. The conditional parameters may access the smart computing devices, such as PDAs, smartphones, and laptops to analyze the nature of various integral components including data management, consumer information system, and mobile workforce management. In addition, the users such as primary user, secondary user, fusion center, and distribution network may access such confidential info from the smart device via security gateway access. From Fig. 10.4, we can observe that the environment has four activities: requesting service, sending location and identity, processing the local authentication, and delivering the services to regulate energy pricing.

10.3.2 Security goals

The major security goals are enlisted to improve key features of smart grid computing systems (Khan, Asif, Ahmad, Alharbi, & Aljuaid, 2020). They are as follows:

- *Proper user authentication*: Since user data are sensitive in nature, the proposed LEPP uses a random session key to ensure the authenticity of the data access.
- *Session-key establishment*: The smart computing device and sensing units share a common session-key to protect the privacy of the data access.
- *No requirement of verification table*: To prevent the key impersonation attack, the system parameters such as user credentials and the secret key are preserved through a dedicated system gateway.
- *User data-freshness*: To ensure data-freshness, the proposed LEPP employs the recent data of the power grid. As a result, the adversaries cannot use the previously requested messages to launch the replay attack.

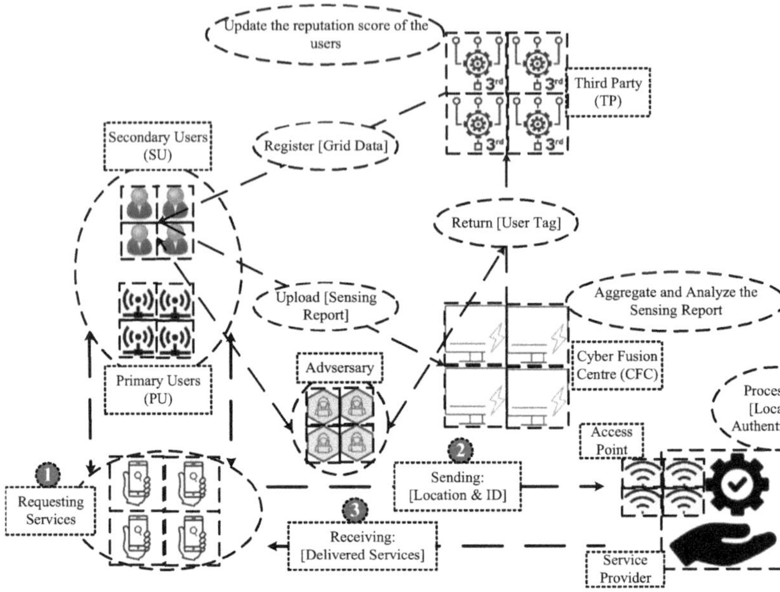

FIGURE 10.4 System architecture of location-based privacy-preserving system.

- *Low communication and computation overhead*: As the sensing unit and smart device have limited resource constraints, a lightweight authentication mechanism is chosen to mitigate the overall cost of the grid computing systems.

10.4 Proposed lightweight efficient privacy-preserving mechanism

The proposed LEPP is composed of three systems phases: system initialization, system authentication, and secret key session. It has two typical scenarios to secure the communication between the control center and the smart breaker. The former scenario assumes the smart breaker as sender-end and the control center as receiver-end to verify the process of authorization and privacy-preserving. Moreover, the latter works in reverse to validate the measurement metrics such as current, voltage, time, and frequency. The detailed descriptions are as follows:

- *Initialization Phase*—This phase generates a unique identity to the smart breaker and accordingly records the breaker identities into the control center. Also, this phase manages the identities of the smart breaker to associate a hashing key H_K via a system gateway.
- *Authentication Phase*—This phase authenticates the exchange of data between the smart breaker and the control center.

- *Secret Key Session Phase*—This phase utilizes mutual-key as a session key to establish data communication between the smart breaker and the control center.

10.4.1 Initialization phase

In this phase, the smart breaker S_B chooses the control center as a trusted party T_P. To establish secure communication, the real-time entities apply a suite of cryptographic algorithms. The suite includes hash function, key encryption, and cryptography procedure to encode or decode the transmitted messages securely. Moreover, in practice, the offline power supplier initiates the identity of the smart breaker as ID_j^{sb} and the control center as ID_{CC} to generate a secret key P_{sk} and secure parameters P_c, that is, for each S_B. Table 10.1 shows the important notation used in the proposed LEPP. Fig. 10.5 shows the proposed lightweight efficient privacy-preserving: initialization phase.

10.4.2 Device authentication phase

The traditional power distribution may stop distributing the workload when any faults occur in the power transmission lines. To deal with faulty status, the smart breaker explores the property of self-healing SH which can communicate the fault with C_C to break the workload. However, SH demands a process of authentication between S_B and C_C to manage the service identities. Upon the confirmation of the workload status, the control center may send its action-taken report to the other S_B. To solve the issues analytically, the action-taken reports are managed through C_C in its dedicated software application. This application has the ability to diagnose the faults of any specific S_B remotely to make a proper decision with minimum power losses. The computation steps are as follows:

Phase–I: The execution steps are as follows:

Step 1: S_B selects an unknown random integer $\rho_i \rightarrow Z^*$ to compute $\mathrm{ID}_{w_j}^{S_B} = \mathrm{ID}_j^{S_B} \oplus \rho_i$

Step 2: S_B utilizes $\mathrm{ID}_{w_j}^{S_B}$ to compute $M_{w_j}^{S_B}$ through the unknown parameter of secure key σ_j: $M_{w_j}^{S_B} = E\left\langle \mathrm{ID}_{w_j}^{S_B}, \sigma_j \right\rangle$

Step 3: S_B uses $H_j^{S_B}$ of $\mathrm{ID}_{w_j}^{S_B}$ to concatenate the system parameter with ρ_i to find the value of ρ_i: $H_j^{S_B} = H\left\langle \mathrm{ID}_{w_j}^{S_B} \| \rho_i \right\rangle$

The sender-end S_E uses S_B to forward the authentication message to CC including $\left\langle H_j^{S_B}, M_{w_j}^{S_B}, ID_{w_j}^{S_B} \right\rangle$

Phase–II: The execution steps are as follows:

Step 1: Upon receiving the authentication message, CC uses $M_{w_j}^{S_B}$ to extract $\mathrm{ID}_{w_j}^{S_B}$: $\mathrm{ID}_{w_j}^{S_B} = D\left\langle M_{w_j}^{S_B}, \rho_i \right\rangle$

Step 2: Accordingly, CC extracts ρ_i from the system parameter, that is, $\mathrm{ID}_{w_j}^{S_B}$: $\rho_i = \mathrm{ID}_{w_j}^{S_B} \oplus \mathrm{ID}_j^{S_B}$

TABLE 10.1 Important notation used in proposed lightweight efficient privacy-preserving.

Parameter	Description
$\mathrm{ID}_j^{S_B}$	j^{th} key identity of S_B
ID_{CC}	Key identity of CC
P_{sk}	j^{th} security parameter of S_B
P_c	A security parameter of CC
$\mathrm{ID}_{w_j}^{S_B}$	j^{th} shadow identity of S_B
$M_{w_j}^{S_B}$	Shadow mask of S_B
ρ_i, σ_j	Random number from S_B and CC
$H_j^{S_B}$	Hash function of S_B
$\mathrm{OH}_j^{S_B}$	Hash function obtained from S_B
$\mathrm{ID}_{cr}^{w_j}$	Identity of control shadow
$M_{cr}^{w_j}$	Control shadow of the mask identity
H_j^{cr}	j^{th} identity of hash function
OH_j^{cr}	j^{th} identity of obtained control function
$\oplus$	Exclsuive operator
$\parallel$	Concateenation function
H	Hash function

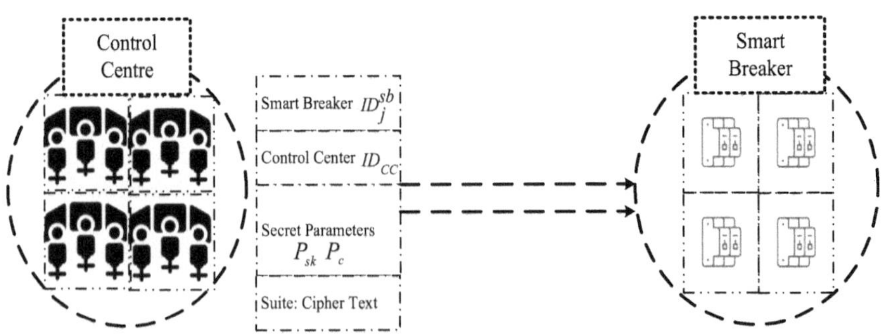

FIGURE 10.5 Proposed lightweight efficient privacy-preserving: initialization phase.

Step 3: *CC* computes $\mathrm{OH}_j^{S_B}$ of $\mathrm{ID}_{w_j}^{S_B}$ to concatenate with ρ_i which confirms the value of ρ_i:$\mathrm{OH}_j^{S_B} = H\left\langle \mathrm{ID}_{w_j}^{S_B} || \rho_i \right\rangle$. To examine the failure status of S_E, $\mathrm{OH}_j^{S_B} = = H_j^{S_B}$ is validated.

Step 4: *CC* generates a random number $\sigma_j \in Z^*$ to calculate the identity of the shadow mask

$\mathrm{ID}_{cr}^{w_j}$:$\mathrm{ID}_{cr}^{w_j} = \mathrm{ID}_j^{S_B} \oplus \sigma_j$

Step 5: *CC* encrypts $\mathrm{ID}_{cr}^{w_j}$ using ρ_j as the unknown key value to compute:

$M_{cr}^{w_j} = E\left\langle \mathrm{ID}_{cr}^{w_j}, \rho_j \right\rangle$

Step 6: *CC* finds the hash value of the control mask $\mathrm{ID}_{cr}^{W_j}$ to concatenate with ρ_j and σ_j. It may compute: $H_j^{cr} = H\left\langle \mathrm{ID}_{cr}^{w_j} \| \rho_j \| \sigma_j \right\rangle$. *CC* sends the message authentication to S_B which involves the parameters including $\left\langle H_j^{\mathrm{cr}}, M_{cr}^{w_j} \right\rangle$

Phase−III: The execution steps are as follows:

Step 1: Upon receiving the message authentication, S_B obtains $\mathrm{ID}_{cr}^{W_j}$ to extract the decryption parameter $\mathrm{ID}_{cr}^{w_j} = D\left\langle M_{cr}^{w_j}, \rho_j \right\rangle$

Step 2: S_B obtains $\mathrm{ID}_{cr}^{W_j}$ to extract the key parameter of σ_j: $\sigma_j = \mathrm{ID}_j^{S_B} \oplus \mathrm{ID}_{cr}^{w_j}$

Step 3: S_B determines the hash value of $\mathrm{ID}_{cr}^{W_j}$ to concatenate with ρ_j to confirm:

$\mathrm{OH}_j^{cr} = H\left\langle \mathrm{ID}_{cr}^{w_j} \| \rho_j \| \sigma_j \right\rangle$. To determine the value of σ_j, $\mathrm{OH}_j^{cr} = = H_j^{cr}$ is validated.

10.4.3 Secret key session phase

The authentication phase successfully passes a mutual symmetric m_k to generate a communication channel to compute: $m_k = H\left\langle \rho_j \| \sigma_j \| \mathrm{ID}_j^{S_B} \right\rangle$.

The existing session achieves better data privacy using m_k. Moreover, the secret parameters such as ρ_j and σ_j to delete the details of previous sessions.

10.5 Security analysis

This section discusses the important key properties of the lightweight authentication mechanism to compare the security efficiencies of smart grid.

Resilience to stolen smart device/medical sensor attack: Assume the smart device is lost or stolen, then A_{dv} may extract the confidential information $\left\langle H_j^{S_B}, M_{w_j}^{S_B}, \mathrm{ID}_{w_j}^{S_B} \right\rangle$ by strategic attacks, such as side-channel/invasive attacks. The tampered data can be used to impersonate as a legal user by A_{dv}. A_{dv} may try to feed the extracted information over a security gateway access, however A_{dv} cannot generate/compute a valid login-request message $\left\langle H_j^{S_B}, M_{w_j}^{S_B}, \mathrm{ID}_{w_j}^{S_B} \right\rangle$. Since $\left\langle H_j^{S_B}, M_{w_j}^{S_B}, \mathrm{ID}_{w_j}^{S_B} \right\rangle$ cannot be determined by A_{dv}, the system parameters such as $\left\langle H_j^{cr}, M_{cr}^{W_j} \right\rangle$ may not be known to infer the user identities. Thus the proposed LEPP can resist a stolen smart device/medical sensor attack.

Resilient to security gateway impersonation attack: Suppose A_{dv} has tampered the confidential information $\left\langle H_j^{\mathrm{cr}}, M_{cr}^{W_j} \right\rangle$ using the attack of node-capture. As a result, A_{dv} can overhear the system authentication message several times, such as $\langle H_j^{S_B}, M_{w_j}^{S_B}, ID_{w_j}^{S_B} \rangle, \langle H_j^{cr}, M_{cr}^{w_j} \rangle$, etc. However, A_{dv} cannot acquire a valid system login-request message $\langle H_j^{cr}, M_{cr}^{w_j} \rangle$ from the previous tampered system login-request message as the value of γ_i changes over S_K. However, A_{dv} cannot impersonate as a security gateway even with the exploitation of node-capture attack.

Resilient to forgery with the node-capture attack: Suppose A_{dv} has tampered with the confidential information $\langle H_j^{cr}, M_{cr}^{w_j} \rangle$ of the user using the attack of node-capture. However, A_{dv} cannot impersonate as a legal user as A_{dv} and cannot produce a valid login request message $\langle \mathrm{ID}_{cr}^{w_j}, \rho_j, \sigma_j \rangle$. This is because $\langle \rho_j \rangle$ cannot be computed until $\langle \sigma_j \rangle$ is determined. Thus the proposed LEPP can be resilient to forgery with the node-capture attack.

Resilient to password-guessing attack: Since the security gateway and smart device do not maintain the password/verifier, the proposed LEPP can resist offline-password-guessing attack.

Resilient to security-gateway bypass attack: Since m_k is not stored in the smart device, the proposed LEPP may resist the attack of security-gateway bypass. Without the knowledge of a security gateway, $\langle H_j^{S_B}, M_{w_j}^{S_B}, \mathrm{ID}_{w_j}^{S_B} \rangle$ cannot be determined. Moreover, the smart devic cannot respond to the user's query, until the security gateway access accepts the system login request. Thus The proposed LEPP can be resilient to security-gateway bypass attack.

Resilient to insider attack: The proposed LEPP initiates a secure hashing key $\langle \rho_j \rangle$ in the system initialization phase. Hence, the security gateway can be resilient to insider attack.

Provision for a zero-knowledge password: Since the proposed LEPP allows only a registered user to authenticate via security gateway access without disclosing the secure session key, the proposed LEPP adheres with the provision of a zero-knowledge password protocol.

Provision of mutual authentication between smartphone/medical sensor and security-gateway access: Since the proposed LEPP prevents forgery attack, it can be more robust in mutual authenticity between smartphone/ medical sensor and security-gateway access.

10.6 Performance analysis

This section discusses a comparison assessments of computation and communication costs to examine a key factor of performance. To effectively analyze, a system phase known as initialization was implemented once. As a result, system authentication was preferred to examine the system costs such as computation and communication. The hardware configuration including 2.5 GHz with Intel i3-M380 was strtegically configured with a dedicated operation system, that is, Windows 10. To implement a logical scenario, a tool known as Matlab was preferred. Moreover, an interoperable language such as Java implemented the logical structure of the cryptosystem.

Table 10.2 shows comparative assessments of computation cost. From Table 10.2, it is observed that during the authentication phase, the computational cost of the proposed LEPP scheme is recorded as <14.3 Ms> whereas the existing schemes (Fouda, Fadlullah, Kato, Lu, & Shen, 2011; Lavanya, Ramaprabha, & Gunaseelan, 2021; Mahmood, Chaudhry, Naqvi, Shon, & Ahmad, 2016; Sule, Katti, & Kavasseri, 2012) are as follows: <24.3 Ms>, <34.63 Ms>, <36.83 Ms>, and <37.43 Ms> respectively.

To examine the communication cost, the proposed LEPP and other existing schemes consider the phase of authentication as $\left\langle H_j^{S_B}, M_{w_j}^{S_B}, \mathrm{ID}_{w_j}^{S_B} \right\rangle$. The key sizes of the system parameters are set to be $<3 \times 128 = 384$ bits$>$ whereas the upcoming requirement is recorded to be $<2 \times 128 = 256$ bits$>$. The examination result reveals that the proposed LEPP can achieve less communication cost than other existing schemes (Fouda et al., 2011; Lavanya et al., 2021; Mahmood et al., 2016; Sule et al., 2012) to improve the performance efficiency (as shown in Table 10.3).

TABLE 10.2 Comparative assessment of computational cost.

Existing schemes	$S_E \rightarrow R_E$	$R_E \rightarrow S_E$	Total cost (ms)
Proposed lightweight efficient privacy-preserving	8.2	6.1	14.3
Mahmood et al. (2016)	13.2	11.1	24.3
Fouda et al. (2011)	18.21	16.42	34.63
Sule et al. (2012)	19.35	17.48	36.83
Lavanya et al. (2021)	18.22	19.21	37.43

TABLE 10.3 Comparative assessment of computational cost.

Existing schemes	Phase I	Phase II	Phase III	Total cost <ms>
Proposed lightweight efficient privacy-preserving	366	258	122	746
Mahmood et al. (2016)	386	262	126	774
Fouda et al. (2011)	406	266	130	802
Sule et al. (2012)	426	270	134	830
Lavanya et al. (2021)	446	274	138	858

10.7 Conclusions

Smart grid computing plays a crucial role in providing computing services such as state estimation, real-time pricing, and periodic billing. These computing services exploit different levels of fine-grained data to analyze metrics, such as data accuracy and metering frequency. The Usecase model of privacy-preserving considers consumer data to manage the energy management issues. To prevent system outages, entities such as smart breaker and control center design a lightweight privacy-preserving mechanism. Moreover, a large-scale blackout seriously demands high-security features to minimize computation and communication costs. To pursue with high-level security, the proposed LEPP utilized a crypto-hash function with control mask identity to explore key features of the symmetric cryptosystem. To analyze the security issues in smart grid communication, this paper explored a strategy of system verification using *REF* 542 *plus* that resolves the problem of time synchronization. The security and performance analysis prove that the proposed LEPP can fulfill all the demands of the sustainable grid computing environment to meet the security level of Industry 4.0.

References

Agarkhed, J., & Ashalatha, R. (2017). A review on challenges and applications of internet of things. *Journal of Innovation in Computer Science and Engineering*, *6*(2), 52–54.

Albayati, A., Abdullah, N. F., Abu-Samah, A., Mutlag, A. H., & Nordin, R. (2020). A serverless advanced metering infrastructure based on fog-edge computing for a smart grid: A comparison study for energy sector in iraq. *Energies*, *13*(20), 5460.

Alsamhi, S. H., Afghah, F., Sahal, R., Hawbani, A., Al-qaness, A. A., Lee, B., & Guizani, M. (2021). Green internet of things using UAVs in B5G networks: A review of applications and strategies. *Ad Hoc Networks*, 102505.

Aman, M. N., Chua, K. C., & Sikdar, B. (2017). Mutual authentication in IoT systems using physical unclonable functions. *IEEE Internet Things of Journal*, *4*, 1327–1340.

Aswathy, G. P., Gopakumar, K., & Ahamed, T. I. (2020). Joint sub-Nyquist wideband spectrum sensing and reliable data transmission for cognitive radio networks over white space. *Digital Signal Processing*, *101*, 102713.

Bakdi, A., Bounoua, W., Guichi, A., & Mekhilef, S. (2021). Real-time fault detection in PV systems under MPPT using PMU and high-frequency multi-sensor data through online PCA-KDE-based multivariate KL divergence. *International Journal of Electrical Power & Energy Systems*, *125*, 106457.

Bohloul, S. M. (2020). Smart cities: A survey on new developments, trends, and opportunities. *Journal of Industrial Integration and Management*, *5*(03), 311–326.

Chaudhari, B. S., Zennaro, M., & Borkar, S. (2020). LPWAN technologies: Emerging application characteristics, requirements, and design considerations. *Future Internet*, *12*(3), 46.

Cioffi, R., Travaglioni, M., Piscitelli, G., Petrillo, A., & De Felice, F. (2020). Artificial intelligence and machine learning applications in smart production: Progress, trends, and directions. *Sustainability*, *12*(2), 492.

Cirani, S., Picone, M., Gonizzi, P., Veltri, L., & Ferrari, G. (2015). IoT-OAS: An oauth-based authorization service architecture for secure services in IoT scenarios. *IEEE Sensors Journal*, *15*, 1224–1234.

Dalby, K. N., Berger, J. A., Brand, H. E., Cairney, J. M., Eder, K., Eggins, S. M., ... Palm, A. B. (2018). Analytical techniques for probing small-scale layers that preserve information on gas–solid interactions. *Reviews in Mineralogy and Geochemistry*, *84*(1), 103–175.

Feng, C., Wang, Y., Chen, Q., Strbac, G., & Kang, C. (2020). Smart grid encounters edge computing: Opportunities and applications. *Advances in Applied Energy*, 100006.

Fouda, M.M., Fadlullah, Z.M., Kato, N., Lu, R., & Shen, X. (2011, April 10–15). Towards a light-weight message authentication mechanism tailored for smart grid communications. In *Proceedings of the 2011 IEEE Conference on Computer Communications Workshops (INFOCOM WKSHPS)*, Shanghai, China (pp. 1018–1023).

Gao, Z., Zhu, H., Liu, Y., Li, M., & Cao, Z. (2013). Location privacy in database-driven cognitive radio networks: Attacks and countermeasures. In *INFOCOM, 2013 Proceedings IEEE* (pp. 2751–2759). IEEE.

Grissa, M., Yavuz, A., & Hamdaoui, B. (2015, December). LPOS: Location privacy for optimal sensing in cognitive radio networks. In *Proceeding IEEE Global Communications Conference (GLOBECOM)*, San Diego, CA, USA (pp. 1–6).

Gupta, M., & Kumar, V. (2020, May). A survey on spectrum sharing techniques in cognitive radio-based smart grids. In *International Conference on Wireless Intelligent and Distributed Environment for Communication* (pp. 113–122). Springer, Cham.

Hajian, G., Ghahfarokhi, B. S., Vasfi, M. A., & Ladani, B. T. (2021). Privacy, trust, and secure rewarding in mobile crowd-sensing based spectrum monitoring. *Journal of Ambient Intelligence and Humanized Computing*, 1–21.

Hummen, R., Shafagh, H., Raza, S., Voig, T., & Wehrle, K. (2014, June 30–July 3). Delegation-based authentication and authorization for the IP-based internet of things. In *Proceedings of the 11th Annual IEEE International Conference on Sensing, Communication, and Networking (SECON)* (pp. 284–292). Singapore.

Jain, S. K., & Kaur, B. (2020). Optimal power transmission for various spectrum sharing approaches in OFDM based cognitive radio network. *ICTACT Journal on Communication Technology*, *11*(03).

Jiang, H., Li, J., Zhao, P., Zeng, F., Xiao, Z., & Iyengar, A. (2021). Location privacy-preserving mechanisms in location-based services: A comprehensive survey. *ACM Computing Surveys (CSUR)*, *54*(1), 1–36.

Jokhio, F., Ashraf, A., Lafond, S., & Lilius, J. (2013, September). A computation and storage trade-off strategy for cost-efficient video transcoding in the cloud. In *2013 39th Euromicro Conference on Software Engineering and Advanced Applications* (pp. 365–372). IEEE.

Keshk, M., Turnbull, B., Sitnikova, E., Vatsalan, D., & Moustafa, N. (2021). Privacy-preserving schemes for safeguarding heterogeneous data sources in cyber-physical systems. *IEEE Access*, *9*, 55077–55097.

Khalfi, B., Hamdaoui, B., Guizani, M., & Elmaghbub, A. (2021). Scalable spectrum database construction mechanisms for efficient wideband spectrum access management. *Physical Communication*, *46*, 101318.

Khan, F. A., Asif, M., Ahmad, A., Alharbi, M., & Aljuaid, H. (2020). Blockchain technology, improvement suggestions, security challenges on smart grid and its application in healthcare for sustainable development. *Sustainable Cities and Society*, *55*, 102018.

Kim, A., Oh, J., Ryu, J., & Lee, K. (2020). A review of insider threat detection approaches with IoT perspective. *IEEE Access*, *8*, 78847–78867.

Lau, F., Rubin, S.H., Smith, M.H., & Trajkovic, L. (2000, October). Distributed denial of service attacks. In *Smc 2000 Conference Proceedings. 2000 IEEE International Conference on Systems, Man and Cybernetics.'Cybernetics Evolving to Systems, Humans, Organizations, and their Complex Interactions'*(cat. no. 0, Vol. 3, pp. 2275–2280). IEEE.

Lavanya, D. L., Ramaprabha, R., & Gunaseelan, K. (2021). Privacy preserving physical layer authentication scheme for LBS based wireless networks. *Defence Science Journal*, *71*(2), 241–247.

Li, S., Zhu, H., Gao, Z., Guan, X., Xing, K., & Shen, X. (2012, March). Location privacy preservation in collaborative spectrum sensing. In Proceeding of IEEE INFOCOM, Orlando, FL, USA, pp. 729–737.

Lowczowski, K., Lorenc, J., Zawodniak, J., & Dombek, G. (2020). Detection and location of Earth fault in MV feeders using screen earthing current measurements. *Energies*, *13*(5), 1293.

Mahmood, K., Chaudhry, S. A., Naqvi, H., Shon, T., & Ahmad, H. F. (2016). A lightweight message authentication scheme for Smart Grid communications in power sector. *Computers & Electrical Engineering*, *52*, 114–124.

Mandal, M. (2020). Privacy-preserving fully anonymous ciphertext policy attribute-based broadcast encryption with constant-size secret keys and fast decryption. *Journal of Information Security and Applications*, *55*, 102666.

Mao, Y., Chen, T., Zhang, Y., Wang, T., & Zhong, S. (2015). Protecting location information in collaborative sensing of cognitive radio networks. In *Proceedings of the 18th ACM International Conference on Modeling, Analysis and Simulation of Wireless and Mobile Systems* (MSWiM), Cancun, Mexico (pp. 219–226).

Mohamed, N., Al-Jaroodi, J., & Jawhar, I. (2020, July). Opportunities and challenges of data-driven cybersecurity for smart cities. In *2020 IEEE Systems Security Symposium (SSS)* (pp. 1–7). IEEE.

Moosavi, S. R., Gia, T. N., Rahmani, A. M., Nigussie, E., Virtanen, S., Isoaho, J., & Tenhunen, H. (2015). SEA: A secure and efficient authentication and authorization architecture for IoT-based healthcare using smart gateways. *Procedia Computer Science*, *52*, 452–459.

Morris, T., Srivastava, A., Reaves, B., Gao, W., Pavurapu, K., & Reddi, R. (2011). A control system testbed to validate critical infrastructure protection concepts. *International Journal of Critical Infrastructure Protection*, *4*(2), 88–103.

Muzammal, M., Talat, R., Sodhro, A. H., & Pirbhulal, S. (2020). A multi-sensor data fusion enabled ensemble approach for medical data from body sensor networks. *Information Fusion*, *53*, 155–164.

Nguyen, D. C., Ding, M., Pham, Q. V., Pathirana, P. N., Le, L. B., Seneviratne, A., ... Poor, H. V. (2021). Federated learning meets blockchain in edge computing: Opportunities and challenges. *IEEE Internet of Things Journal*, *8*(16), 12806–12825.

Nicolas, K., & Wang, Y. (2019, October). A novel double spending attack countermeasure in blockchain. In *2019 IEEE 10th Annual Ubiquitous Computing, Electronics & Mobile Communication Conference (UEMCON)* (pp. 0383–0388). IEEE.

Padmavathi, D.G., & Shanmugapriya, M. (2009). A survey of attacks, security mechanisms and challenges in wireless sensor networks. arXiv preprint arXiv:0909.0576.

Pan, H., Dou, Z., Cai, Y., Li, W., Lei, X., & Han, D. (2020, September). Digital twin and its application in power system. In *2020 5th International Conference on Power and Renewable Energy (ICPRE)* (pp. 21–26). IEEE.

Perelmuter, G. (2021). Present future: Business, science, and the deep tech revolution. Greenleaf Book Group.

Rahouti, M., Xiong, K., & Xin, Y. (2020). Secure software-defined networking communication systems for smart cities: Current status, challenges, and trends. *IEEE Access*, *9*, 12083–12113.

Rid, T., & Buchanan, B. (2015). Attributing cyber attacks. *Journal of Strategic Studies*, *38* (1–2), 4–37.

Ropp, M.E., Reno, M.J., Bower, W., Reilly, J., & Venkata, S.S. (2020). *Secondary networks and protection: Implications for DER and microgrid interconnection* (No. SAND2020−11209). Albuquerque, NM (United States): Sandia National Lab.(SNL-NM).

Shao, J., & Zhang, J. (2020). Communication-computation trade-off in resource-constrained edge inference. *IEEE Communications Magazine*, *58*(12), 20−26.

Shi, E., Chan, T.H., Rieffel, E.G., Chow, R., & Song, D. (2011). Privacypreserving aggregation of time-series data. In *Proceeding Network Distributed System Security Symposium (NDSS)*, San Diego, CA, USA.

Sikder, A.K. (2020). A comprehensive security framework for securing sensors in smart devices and applications, Available from: https://doi.org/10.13140/RG.2.2.35553.02409.

Singhal, S., Gankotiya, A.K., Agarwal, S., & Verma, T. (2012, January). An investigation of wireless sensor network: A distributed approach in smart environment. In *2012 Second International Conference on Advanced Computing & Communication Technologies* (pp. 522−529). IEEE.

Sule, R., Katti, R.S., & Kavasseri, R.G. (2012, July 22−26). A variable length fast message authentication code for secure communication in smart grids. In *Proceedings of the 2012 IEEE Power and Energy Society General Meeting*, San Diego, CA, USA (pp. 1−6).

Tashman, D. H., & Hamouda, W. (2020). An overview and future directions on physical-layer security for cognitive radio networks. *IEEE Network*, *35*(3), 205−211.

Tewari, A., & Gupta, B. B. (2020). Security, privacy and trust of different layers in Internet-of-Things (IoTs) framework. *Future Generation Computer Systems*, *108*, 909−920.

Trostle, J., & Parrish, A. (2010). Efficient computationally private information retrieval from anonymity or trapdoor groups. In *International Conference on Information Security* (pp. 114−128). Springer.

Vinodh, S., Antony, J., Agrawal, R., & Douglas, J. A. (2020). Integration of continuous improvement strategies with Industry 4.0: A systematic review and agenda for further research. *The TQM Journal*, *33*(2), 441−472.

Wu, G., Mizuno, M., & Havinga, P. J. (2002). MIRAI architecture for heterogeneous network. *IEEE Communications Magazine*, *40*(2), 126−134.

Zhang, L., Fang, C., Li, Y., Zhu, H., & Dong, M. (2015). Optimal strategies for defending location inference attack in database-driven crns. In *Communications (ICC), 2015 IEEE International Conference on*.

Zhang, Z., Zhang, H., He, S., & Cheng, P. (2015). Achieving bilateral utility maximization and location privacy preservation in database-driven cognitive radio networks. In *Mobile Ad Hoc and Sensor Systems (MASS), 2015 IEEE 12th International Conference on*. IEEE.

Chapter 11

Smart grid metering: security, privacy, and open challenges

P. Muralidhara Rao[1], Fadi Al-Turjman[2] and B.D. Deebak[1]
[1]School of Computer Science and Engineering, Vellore Institute of Technology, Vellore, India,
[2]Department of Artificial Intelligence Engineering, Research Center for AI and IoT, AI and Robotics Institute, Near East University, Mersin, Turkey

11.1 Introduction

The energy grid's operation is built on a top-down basis, centralized producer-to-consumer power generation and distribution system. Electricity is generated in this process by (1) huge generators that are mostly fossil/nuclear-fueled and are centrally located (e.g., nuclear and coal-fired power plants). Productivity is changed based on the grid's current (e.g., electricity demand) and projected (e.g., the demand forecasts for the next day). The majority of these manufacturers are located in rural areas. They generate high-voltage electricity to transport the generated electricity to consumers, which can then be (2) transferred over vast distances via transmission lines. (3) Transformers can be used to reduce the voltage of electricity once it reaches the customers' area. (e.g., cities, industrial districts), making it useful for the customers' equipment (Kothari & Nagrath, 2003). It is worth noting that this stepping-down procedure might happen numerous times, changing based on the needs of consumers, where the voltage should be set at a certain level in the transformer's proximity (different voltage levels are required for industrial equipment than for household appliances). Finally, four distribution lines can transmit power to five users, which is used to meet device demand (Arif, 2013). In a simplified example, this procedure is depicted in Fig. 11.1.

Human involvement in the energy grid's specified supply chain's resilient operation is widespread. The different controls and monitoring systems for the creation of power are not automated and require manual intervention. As a result, specialized individuals who are accountable for the proper operation of these processes are necessary. Furthermore, many personnel are involved in the supply chain's various transit and transformation activities. People in charge of servicing transmission lines and transformers, as well as employees

Sustainable Networks in Smart Grid. DOI: https://doi.org/10.1016/B978-0-323-85626-3.00009-0

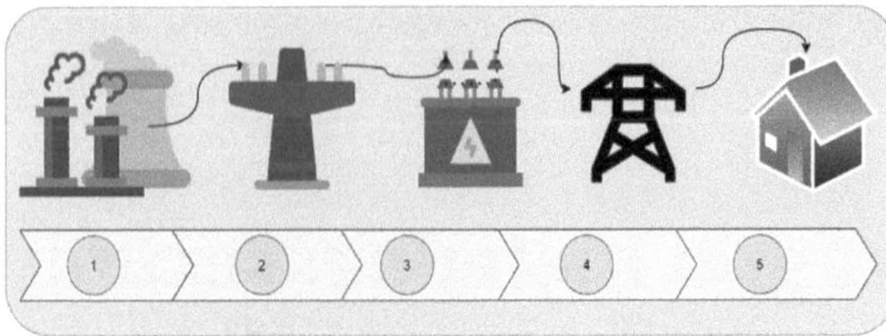

FIGURE 11.1 A hierarchical electric transformation.

in charge of electricity distribution and monitoring throughout various supply chain sectors (Xu, Yu, Griffith, & Golmie, 2018).

Energy system transitions to future SG are difficult with respect to politics, economics, law, and technology. While many people believe that parts of life are progressing at a rate that would have been unfathomable a previous generation, one aspect has mostly remained dormant: People who use electricity. So far, involvement of customers has been two questions that characterize it: "Should I get the environmentally friendly appliance?" "Should I buy the eco-friendly appliance?" and "Should I buy the eco-friendly appliance?," "Should I get the eco-friendly appliance?" "Should I buy the eco-friendly appliance?" and "Should I buy the eco-friendly appliance?" "Is solar going to be a good investment for me?" Consumers' social and psychological factors, on the other hand, can have a significant impact on the resilience of SG. In the SG, key management is a significant security concern for maintaining data confidentiality and integrity.

Many key management solutions have been developed to meet these needs; However, many have flaws such as a lack of flexibility, scalability, and message support communication. In May 2010, According to a report provided by krebson-security, the FBI examined the hacking of smart grid meters in Puerto Rico. An intelligence warning was disseminated to select industry personnel and law enforcement officials regarding the bureau's findings. Because of "the attacker's and electric user's ease of exploitation and economic gain," the FBI suspects that former meter makers and utility employees modified the meters in exchange for cash and training others to do so. As a result, it is critical to ensure that information transmitted by SG technologies is kept private and accessible only to the intended receivers (Gungor & Lambert, 2006). Given the SG's immense scale and sophisticated architecture, it is easy to see how the vulnerabilities connected with the smart-grid communication system may be enormous.

These security flaws must be solved for smart grids to be secure and perform effectively, optimize adoption, and successfully deliver on the promise of SG investment. Although governments, businesses, and academia have defined most of the architectures, frameworks, and roadmaps for SG, there are still several major security and privacy challenges in SG communications. These concerns are now being explored. The SG is a next-generation electricity system gaining traction among government, industry, and academic institutions. It's a modernized electrical grid that relies on two-way digital connections between the supplier and the consumer to support intelligent metering and monitoring devices. Given the growing importance of energy utilities, SG technology introduces new security dangers into our daily lives that must be addressed.

Grid instability, utility fraud, and the theft of customer information and energy usage data might all result from deploying an SG without sufficient protection. Because of the various communication architectures of smart grids, developing sophisticated and powerful security mechanisms that can be easily applied to protect communications between different layers of the smart grid is a challenging infrastructure problem (Al-Badi, Ahshan, Hosseinzadeh, Ghorbani, & Hossain, 2020). This chapter looks at the communication security of an SG from the point of view of metering and control of cryptographic approaches. Also, we reviewed various mechanisms to analyze threats to system security, risks and theft of services, and privacy threats are all included in this threat taxonomy.

11.2 Smart grid: an overview

An intelligent grid, also known as a smart power grid, is a significant upgrade to the standard power system of the twentieth century. It is an electrical system that integrates information, two-way, cyber-secure communication technologies, and computational intelligence to achieve a clean, safe, secure, reliable, resilient, efficient, and sustainable system across electricity generation, transmission, distribution, and consumption (Fang, Misra, Xue, & Yang, 2011). The original aim of SG, on the other hand, was to leverage advanced metering infrastructure (AMI) to increase demand-side management, energy efficiency, and reliable grid protection against deliberate sabotage (Rahimi & Ipakchi, 2010). Due to increased demand for its services, it has been expanded.

The Energy Independence and Security Act of 2007 was signed into law in 2007 mandated that the NIST oversee R&D of a framework to enable SG system and device interoperability. Although a specific and thorough definition of SG has yet to be presented, the following are the predicted benefits and needs of SG, according to the NIST report. Despite the lack of a specific and complete definition of SG, t NIST (Arnold et al., 2010) has predicted the following benefits and requirements for future smart grid systems:

1. Improving power quality and dependability, as well as optimizing facility utilization, to avoid the need for backup power plants;
2. Increasing the capacity and efficiency of current power grids;
3. Increasing system resilience and allowing for proactive maintenance and self-healing responses to system failures;

4. Distributed power sources can be accommodated, and renewable energy sources can be deployed more widely.
5. Automating maintenance and operations;
6. Electrification of cars and the development of new energy sources to reduce carbon gas emissions;
7. Lowering consumption of oil by eliminating the requirements for wasteful generation during peak demand periods;
8. Offering possibilities for improved grid security and facilitating the transition to plug-in electric automobiles and new energy storage choices;
9. Expanding the number of options and interactions available to consumers, as well as enabling new goods, services, and markets.

NIST presented a conceptual model, as shown in Fig. 11.2, to actualize this new grid paradigm. It could be used as a starting point for the different

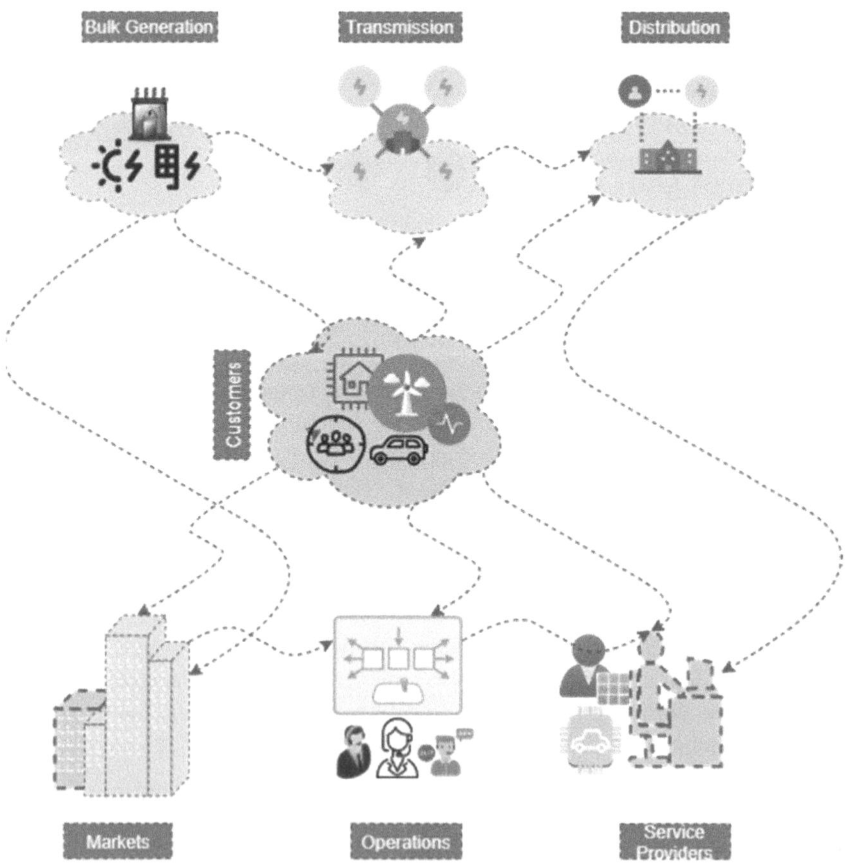

FIGURE 11.2 Smart grid conceptual model-based National Institute of Standards and Technology standard (Arnold et al., 2010).

components of the electric system that are undergoing work on SG standardization. The SG is divided into seven domains in this conceptual approach. Each domain has an SG actor or a group of SG actors, like devices, systems, or programs, that make decisions and transmit data in order to complete tasks. Table 11.1 shows the basic descriptions of the domains and actors. It's also worth noting that NIST presented this model from a variety of perspectives of the various roles participating in the SG (Arnold et al., 2010; Pérez-Guzmán, Salgueiro-Sicilia, & Rivera, 2017).

Most countries have invested in SG research and development. The SG, for example, is an important part of The American Recovery and Investment Act, which President Obama signed into law, and includes $11 billion in financing to "jump-start the transformation to a larger, better, SG." One of the driving forces behind the United States' current extensive work program toward SG is the necessity to update its electricity system. Under-investment in infrastructure and rising energy consumption resulted in a insufficient electricity distribution network resilience under stress conditions. In addition, there are rising expectations for integrating a variety of renewable energy sources into the power system. As a result, the US Department of Energy's SG program for R&D (DoE, 2010) has established performance objectives for 2030: a reduction of 20% in peak energy demand; 100% accessibility for all key loads all of the time, as well as a variety of dependable services for other loads; and a 40% increase to attain a load factor of 70% and improve system efficiency and asset utilization (200 GW). There are several large-scale industrial research initiatives now active, such as IBM GridWise (Sundstrom, Gantenbein, & Binding, 2011) (Fig. 11.3).

TABLE 11.1 Actors and domains involved in smart grid paradigm.

Domain specification	Actors involved in the specified domain
Service provider(s)	• Organizations that provide electric customers and utilities with services
Global and local markets	• The electrical market's operators and participants
Customer(s)	• The electrical utility's end users. Energy resources can also be generated, managed, and stored by actors.
Bulk production	• It can generate large amounts of electricity in bulk and store and generate electricity for future operations.
Electricity transmission	• Long-distance transmitters of massive amounts of energy. • It also can generate and store electricity for later use.
Electricity distribution	• To manage the transmission of power between consumers' homes. • It can also produce and store electricity for later use.

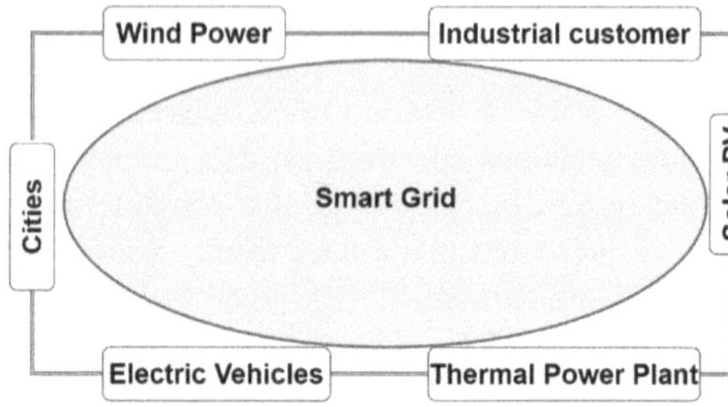

FIGURE 11.3 Typical smart grid.

There is already an EU-wide program of major expenditures in modernizing the distribution network, reducing the pressure for quick decisions on disruptive new technology adoption (Giordano & Fulli, 2012). The most pressing challenge in Europe is integrating renewable energy generation to reach the 2020 CO_2 emission reduction targets from the production of fossil fuels. Because of the variable nature of these energy sources, conventional transmission and distribution networks are unable to meet these needs. As to become less reliant on traditional and nonconventional sources of energy, a significant amount of effort will be required. SGs are emphasized as a crucial application in the European Commission's 2020 policy document (Delarue, 2009), in the European technology platform of SG (Fan et al., 2012), as well as the upcoming Future Extensive Research Program [FI PPP, (European Commission, 2010)]. FINSENY, led by Siemens, is a recent example of a FI PPP initiative on an SG. The EU has devised a carefully planned approach to introducing SG technologies in the medium to long term through the technological development platform. The priority is establishing standards, research initiatives involving academics and industry (utilities and manufacturers), and demonstration/pilot projects.

11.3 Smart grid metering communication technologies

11.3.1 Home area network

The HAN is a network that is used to track and control energy use inside the home. It connects to the utility's electric meter and tracks energy usage, enabling the homeowner to see how much energy is used and where it is used. It is also used to provide the utility the ability to control heating and air conditioning to save energy (https://www.energy.gov/sites/prod/files/gcprod/documents/Smart_Grid_Communications_Requirements). The HAN also makes it simple to turn appliances on and off. The HAN may use wireless or AC power line communications. The HAN is the local arm of the national SG energy conservation initiative (https://indiasmartgrid.org/Home-Area-Network.php). The Home Area Market is already forming to offer homes with digital application solutions and energy management within the smart grid sector. Because utilities are looking for ways to adopt demand-side management programs, the HAN has ramifications for homeowners, utilities, and smart grid providers. From 1.5 million in 2009 to 14.7 million in 2014, the number of dwellings has increased dramatically, the global Smart home networks have a large installed base that is predicted to grow (Akkermans et al., 1999).

The HAN is a Smart Grid subsystem dedicated to DSM that incorporates energy efficiency and demand response. These are critical components in the creation of value from a SG deployment. The following are a few applications of distributed generation:

- Energy conservation through behavior
- Technology-assisted dynamic pricing
- *Deterministic direct* load control.

These applications can be enabled by high-bandwidth, two-way, end-to-end connections in the Smart Grid. A Smart Grid that incorporates energy efficiency and demand response increases its worth as a long-term infrastructure investment while also shortening the time it takes to achieve a satisfactory short-term return on investment.

11.3.2 Meter local area network

The communication between the smart meter and the data concentrator is referred to as the meter local area network. The MLAN network is primarily found in two places: (1) the energy distribution domain, which includes data concentrators, and (2) the field area networks, which include many automated devices such as monitors, re-closers, switches, capacitor controllers, and so on. MLAN commonly communicates via powerline communication (PLC), which sends data from one node to another via existing wireline connections. Narrowband PLC (NB-PLC) and broadband PLC are two types of PLCs that operate at various data rates (BB-PLC).

11.3.3 Wide area network

To convey information, wide area networks (WAN) establish communication links between NANs and utility systems. An effective, stable, and secure communication network, as one of the basic technologies, is critical to achieving all of the aims of smart grid NANs. The enterprise network is extended to remote places, with data from access networks being returned to the enterprise. Broadband systems, including wired and wireless, point-to-point and point-to-multipoint, fiber and microwave are all covered. All Smart Grid access networks will be built on top of these backhaul networks.

11.4 Smart grid metering: security and privacy

Security mechanisms should be upgraded from both a physical and a cyber perspective to increase the grid's sturdiness and reliability. In the end, this measure will lessen the likelihood and effects of artificial disasters. security of energy is a concept that ensures the availability of energy source reliability, ensures an adequate energy supply at a reasonable price, and prevents negative environmental consequences. Risk management, energy diversification, and policy implementation decision-making are all aspects of energy security. SG technologies integrating renewable energy can improve the electric system's energy security and safety (Finster, Phillips, & Wallace, 2016). An SG is a solution for modernizing the infrastructure and electrical energy system to present a more intelligent and dependable power grid. SGs have numerous advantages over traditional grids. SGs increase grid system dependability and sustainability by improving both physical and economic operations (Al-Turjman & Abujubbeh, 2019).

Although the SG offers numerous social and technical advantages, it also raises numerous security and privacy problems. Customers work closely with utilities to monitor energy usage in the smart grid, which necessitates them sharing more information about their energy consumption and exposes them to privacy breaches. Because they are linked by a vast network of computerized meters and equipment, grid customers and infrastructure are vulnerable to scalable network-based attacks. We examine security and privacy vulnerabilities due to this initiatives and new infrastructure that could help decrease coverage to these negative consequences. The smart meter is a usage-reporting gadget installed at each client location. It is a computerized version of a power meter found outside many of today's homes (Patel et al., 2017). Non-volatile storage, a processor, and communication capabilities are all included in each smart meter.

Although the smart meter's appearance and operation are similar to its less sophisticated predecessor in many ways. It has many features that make it more handy. Smart meters can keep track of how much energy is used at different times of the day, remotely disconnect a customer, or In the event of an issue, send out alarms. The smart meter can also control "smart" appliances directly, such as turning down the air conditioner during peak hours. Smart meters also have unanticipated privacy implications for customers. Energy usage data, The information-rich side channel, which is retained at the meter and later distributed, reveals client habits and behaviors (Quinn, 2009). Power consumption signatures can be detected in certain activities, such as watching television. History has shown that when financial and political incentives are aligned, good things happen, behavioral data mining techniques adapt swiftly to meet the needs of those who want to profit from it.

Utility corporations are not the only companies who can be violating your privacy. For example, the recently introduced Google PowerMeter service deceives real-time usage statistics from installed devices. Even though services like Google PowerMeter is an opt-in service, customers have fewer options for control utility companies' use of electricity data. In general, existing privacy laws in the United States are an assortment of rules and regulations. It is unclear how these, or any other legislation, will affect how customers use energy.

The following are the primary issues that must be overcome to secure the smart grid:

Identity management and access control: It's vital to keep data sent via smart grids private and ensure that only the intended receiver sees the message. Furthermore, the smart grid is composed of many interconnected components. Due to security considerations, authentication is required to validate the receiver's identity to prevent disruption or exploitation. Only authenticated users, organizations, and services access the control center, transmission, and distribution grids (Yan, Qian, Sharif, & Tipper, 2012).
Security and privacy policies: Although adopting regulations on security and privacy should not result in poor latency, there is a large need for appropriate security standards to develop interactions among consumers, utilities, and third parties (Simmhan, Kumbhare, Cao, & Prasanna, 2011).
Threat defense: Because target smart grids are inherently vulnerable, securing them from identified threats is vital by constructing an effective, layered defense system that operates across the grid infrastructure. As to protect against denial-of-service attacks, threat defense uses network segmentation and access control (DoS). It also has several security features, such as firewalls, VPN, and IPS (Gunduz & Das, 2020).
Physical security: SG systems can include hundreds of thousands, if not millions, of distant points and field area networks. Besides, it makes maintaining the SG's physical security difficult. Because of the systems' geographical spread, it may be difficult to reach all terminals for maintenance (Boonzaier, 2006) (Fig. 11.4).
Connectivity: In SGs, communications connectivity entails a move to a distributed system akin to the Internet, in which large numbers of devices are networked. As this is one of the emerging issues in this field, protective procedures are critical (Rivas & Abrao, 2020).

Numerous key management strategies (Deng & Gao, 2021; Khan, Kumar, Ahmad, & Rana, 2021; Odelu, Das, Wazid, & Conti, 2016; Qi & Chen, 2020; Yan, Chang, & Zhang, 2017) for smart grid and SCADA systems have been proposed to accomplish end-to-end secure communication. These schemes, however, have flaws such as a lack of real-time constraints, scalability, and the capacity to handle message exchange in a variety of contexts. A strong and efficient key management solution is necessary to provide high confidentiality in a SG. We describe and explore existing key management solutions for smart grid and SCADA systems, as well as their shortcomings in this chapter.

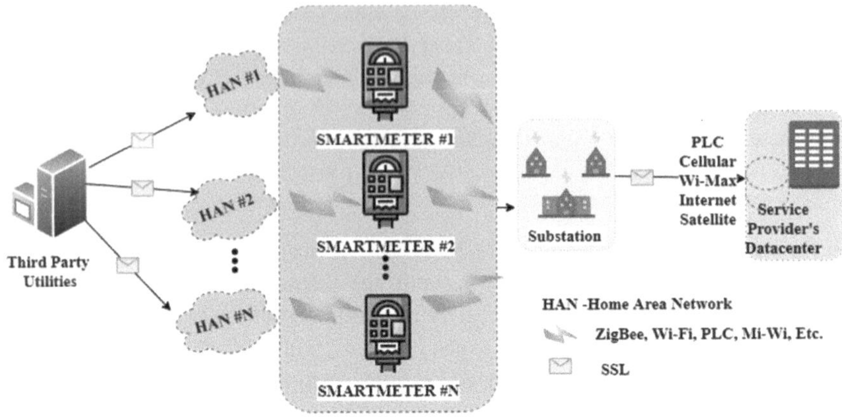

FIGURE 11.4 Smart grid communication and control.

11.5 Security attacks based on smart metering networks

As the SG takes shape, information technology (IT) and telecommunications infrastructures have become critical to the energy sector's infrastructure transforms the converting the electric grid into a two-way flow of electricity and data. As a result, a more diverse energy industry must address the administration and protection of these infrastructures' systems and components. Existing cybersecurity standards and assessment programs exist in the IT and telecommunications sectors to address vulnerabilities and detect known flaws in these systems. Attackers with various intentions and talents can take advantage of the security flaws in the smart grid system and wreak varying degrees of damage. Online hackers, terrorists, laborers, opponents, or clients are examples of top-level attackers. Clients can update data and information, as well as gain authority, without having to pay. Professionals and experts discovered only nine security flaws between 2005 and the beginning of 2010 (Amin & Giacomoni, 2011). The matter has piqued the curiosity of security specialists. As a result, by the end of 2011, 64 vulnerabilities had been discovered. In addition, in the first eight months of 2012, information regarding 98 security flaws was made public (Kesan & Hayes, 2016).

Based on the adversary's goal, we can divide prospective smart grid cyber-attacks into three categories: attacks against devices (such as smart meters), attacks against masters (such as device manufacturers), and communication between devices and masters is under threat.

11.5.1 Security attacks on devices

For a variety of reasons, a potential assailant will find you an intriguing target. To begin with, many of the devices will have intrinsic value due to their simple purpose. A smart meter keeps track of our energy usage, which could be valuable to people who want to harm us. Based on an energy usage model, it decreases at different times of the day. For example, a thief may use this information to figure out when a homeowner will be gone. Smart meters are linked to other devices in the home in some smart grid implementations, allowing for more granular control over energy usage. Allowing a utility to make minor modifications to a home's air conditioning to prevent a blackout might be as simple as that. Unfortunately, hackers may utilize the same approach for less than noble goals. Finally, the value of devices is determined by the information entrusted to them. To be exact and reliable, the SG relies on data from linked meters. Hackers could alter a single meter to reduce a customer's energy cost or conceal more illicit activity, such as narcotics production (Clastres, 2011).

11.5.2 Attacks on communications

Monitoring and altering messages as they are transmitted is a common attack method. These assaults are extremely dangerous due to the volume and sensitivity of data traveling through the smart grid, as messages and data can be intercepted, recorded, or manipulated while in transit (Fig. 11.5).

All of these threats jeopardize the integrity of the information and data being carried, as well as the broader infrastructure. For instance, providing your utility supplier with information on your house or business's energy consumption exposes you to many security risks. An adversary may keep track of your energy usage to see if you experience any outages or downtime at your home or business to organize an assault to name a few examples, a hacker may change the data transmitted to the utility supplier and alter the information (Clastres, 2011).

11.5.3 Attacks on the entire environment

The companies at the heart of the SG will be targeted by a skillful attacker, because a successful attack on one of these network links has the highest chance of causing the most damage—or profit—where device makers, solution providers, network operators, power plants, and even utilities themselves could be targeted. These firms are entrusted with gathering and analyzing sensitive information that attackers try to steal (Clastres, 2011).

11.6 Security goals and open challenges

The primary security goals and open challenges related to smart grids are briefly described in this section.

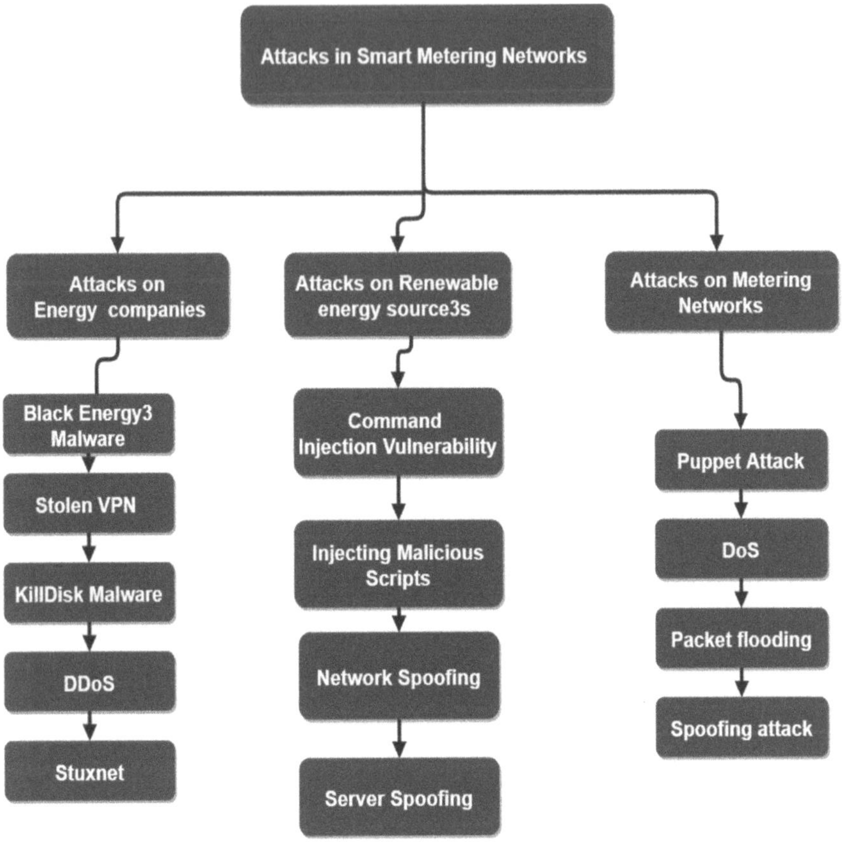

FIGURE 11.5 Attack classification based on smart metering networks.

11.6.1 Data confidentiality

Massive volumes of data are generated by the smart grid, which must be safely received, stored, and analyzed. Some information will be sensitive, such as personal information about customers and the utilities themselves. As to prevent unauthorized disclosure of sensitive information, precautions must be taken. The data shared and stored in SGs must be protected in terms of data confidentiality. A utility company, the main data-generating component in an SG, gathers data from smart gadgets to provide information on energy usage in a home.

In most cases, an energy supplier offers use tariff information, which is open to the public. Billing information is also considered confidential because it is calculated as energy use and usage tariff information (Flick & Morehouse, 2010). The scientific challenge here is to figure out how to keep data about energy usage safe Smart meter data can also be utilized to generate energy usage estimates. Such projections may all actors in the delivery chain will profit (GenCos, TSOs, and DSOs) in terms of implementing load balancing activities that are taken ahead of time. This information can be used by ESCOs along with market pricing estimates, to schedule smart appliances and reduce energy use. Third-party companies can provide forecasting services if they are given access to smart meter data.

11.6.2 Security and privacy

Based on the adversary's goal, we can divide prospective smart grid cyberattacks into three categories: attacks against devices (such as smart meters), attacks against masters (such as device manufacturers), and communication between devices and masters is under threat. The characteristic or state of being separated from the presence or view of others is defined as privacy. Consumer privacy in the context of smart grids involves avoiding disclosing personal information to anyone who's not a consented entity Private data includes information such as a person's name, address, and energy consumption. An energy supplier in a SG needs aggregated data from their to the region determine per-region consumption at a given moment (Franceschetti & Grossi, 2011). It is possible to transfer information with energy providers. An energy distributor's load balancing module takes this information into account when distributing energy as needed.

11.6.3 Data usage

By default, a customer agrees to provide information for energy suppliers with personal information with information regarding billing and aggregated energy consumption. However, regulations including the EU Data Protection Directive must be followed before processing data for non-billing and aggregated energy use purposes (Tene & Wolf, 2013) demands express consumer agreement. Both ESCOs and data users in a smart grid system can access only, customers' affirmative consent is required for purposes that can be revoked at any moment (Schwartz, 2010). Data collection and utilization without explicit consent may pose major security risks. Even if you grant your consent [as considered in (Asghar & Russello, 2011)], it is still a work in progress to guarantee that the data is processed following the consent granted. This unsolved issue is inextricably linked to the smart grid.

11.6.4 Fine-grained access control

Because of the noise and interferences in smart grids, data access control is a difficult problem. On the one hand, fine-grained data access management is critical since unauthorized access to sensitive data can have terrible consequences and be unlawful. On the other hand, access control fault tolerance is critical due to the potential consequences (implied by faults) that could be far more devastating than those involving generic data. Control bits corruption, in particular, could render the security action useless. Another issue is that because the enforcing entity (ESCOs) is semi-trusted, access controls may leak sensitive information about private data (say honest-but-curious). It becomes a problem to impose access controls without revealing the information they are supposed to safeguard (Rizwan Asghar, Ion, Russello, & Crispo, 2013)

11.6.5 Trust

Trust is a fundamental problem in the operational dimension of socio-technical systems, and it is heavily influenced by time. Trust, according to Simmel's "money philosophy," comprises a "quasi-religious faith" that is required to deal with future non-knowledge (Möllering, 2001). People with trust might "act as though the future was guaranteed" (Luhmann, 2018). The temporal dimension of Trust is highlighted in current research on "trusting process perspectives" (Bachmann & Zaheer, 2013). Trust and time have a long history together. People must commit to a "leap of faith" (Möllering, 2006) when developing steady expectations of future changes, and this shift has become a hot topic in the field of trust study.

11.6.6 Transparent auditing verifiability

Consumers in Singapore should be able to ensure that they are not being overcharged and that they are being charged correctly for the amount of energy they consume (duration) and when they spend it (peak or off-peak). On the one hand, energy providers must ensure that customers are charged for the energy they consume and when they consume it. Consumers are rewarded financially if they report lesser consumption and replace peak-hour consumption with off-peak-hour consumption. Energy suppliers, on the other hand, may have a financial incentive to do the exact opposite, that is, to portray more consumption than is actually occurring and to substitute peak-hour usage for off-peak-hour usage.

11.6.7 Availability

The SG's ability to provide fast and, it is vital to have consistent access to and use of information. The benefits of SG cannot be realized without access to actionable, real-time, and trustworthy data. It is critical to consider how data is gathered, processed, and shared, and security solutions must fulfill these objectives while minimizing the impact on availability (Kumari & Tanwar, 2020).

11.6.8 Integrity

The SG is reliant on accurate and reliable data. As to avoid fraud and other more dangerous threats, it is necessary to guarantee that data is reliable and modification-free (Belapurkar et al., 2009).

11.6.9 Tamper resistant

Tamper resistance refers to a product's, package's, or system's resistance to tampering (intended malfunction or sabotage) by either normal users or others with physical access to it. Tamper-resistant mechanisms or features are commonly found on packages to prevent tampering with SG services. Tamper detection, for example, is a problem with smart meters, industrial control systems, and other applications. When the lid of a system is opened, a switch is typically utilized to detect it (Grand, 2005).

11.6.10 Nonrepudiation

Assurance that the transmitter of information receives proof of delivery, and that the recipient receives verification of the sender's identity so that neither party can dispute processing the data. The lack of nonrepudiation is a fundamental roadblock to establishing a reliable smart grid. Bills are created in modern power systems based on the amount of service consumed by residential or commercial customers. Meter readings, on the other hand, may be unreliable owing to malicious behavior (e.g., energy theft) or external threats. The main cause is that, there is no other way for the power provider to get the reading value but to ask users for it.

11.6.11 Accountability

Users of a system should be held accountable for their activities, according to the concept of accountability. It means that all user interactions with sensitive systems should be documented and traced back to a single person. It should be tough to tamper with these logs with and should have a high level of integrity protection.

There is still a lot of research being done to address specific concerns, but no work has been done to address the basic issues in terms of security and privacy associated with SG. In summary, the open challenge is to handle several existing device, service, and communication related issues. In a comprehensive manner, the above research challenges, rather than piecemeal, result in the development of a safe data management framework that can span the complete life-cycle of energy data.

11.7 Conclusions

Despite SMI's many advantages, the large number of heterogeneous devices projected to be coordinated via public communication networks has raised security and privacy concerns. Due to their Internet and wireless networks, SGs, on the other hand, are more vulnerable to security attacks than traditional grids. As a result, this chapter provides a quick review of real cyberattacks in traditional energy networks as well as those aimed at the smart metering network. This chapter proposes a threat taxonomy that takes into account: (1) system-level security concerns; (2) service threats and theft; and (3) privacy threats. A set of security and privacy requirements has been developed based on the dangers that have been highlighted.

References

Akkermans, H., Ottosson, H., Gebremedhin, A., Larsson, L., Karlsson, B. G., Larsson, R., ... Björsell, N. (1999). The KEES project: Energy efficiency in a deregulated market. Malmo, EnerSearch AB.

Al-Badi, A. H., Ahshan, R., Hosseinzadeh, N., Ghorbani, R., & Hossain, E. (2020). Survey of smart grid concepts and technological demonstrations worldwide emphasizing on the Oman perspective. *Applied System Innovation*, *3*(1), 5.

Al-Turjman, F., & Abujubbeh, M. (2019). IoT-enabled smart grid via SM: An overview. *Future Generation Computer Systems*, *96*, 579–590.

Amin, S. M., & Giacomoni, A. M. (2011). Smart grid, safe grid. *IEEE Power and Energy Magazine*, *10*(1), 33–40.

Arif, M. T. (2013). Energy storage and its strategic impacts on the power network (Doctoral dissertation, Central Queensland University Australia).

Arnold, G. W., Wollman, D., FitzPatrick, G., Prochaska, D., Holmberg, D. G., Su, D. H., ... Boynton, P. A. (2010). NIST framework and roadmap for smart grid interoperability standards, release 1.0.

Asghar, M. R., & Russello, G. (2011). *Flexible and dynamic consent-capturing. International workshop on open problems in network security* (pp. 119–131). Berlin, Heidelberg: Springer, June.

Bachmann, R., & Zaheer, A. (Eds.), (2013). *Handbook of advances in trust research*. Edward Elgar Publishing.

Belapurkar, A., Chakrabarti, A., Ponnapalli, H., Varadarajan, N., Padmanabhuni, S., & Sundarrajan, S. (2009). *Distributed systems security: Issues, processes and solutions*. John Wiley & Sons.

Boonzaier, C. G. (2006). Impact of the identification and survey of the administrative area boundaries process on the implementation of the communal land rights act: A case study of the Eastern Cape Province (Doctoral dissertation).

Clastres, C. (2011). Smart grids: Another step towards competition, energy security and climate change objectives. *Energy Policy*, *39*(9), 5399–5408.

Delarue, R. (2009). The ILO and its role in responding to the financial, economic and social crisis, promoting decent work and strengthening the social dimension of globalisation. *Social developments in the European Union*, 65.

Deng, L., & Gao, R. (2021). Certificateless two-party authenticated key agreement scheme for smart grid. *Information Sciences*, *543*, 143–156.

DoE, U. S. (2010). Smart grid research & development: Multi-year program plan (mypp) 2010–2014. US Department of Energy, Office of Electricity Delivery & Energy Reliability.

European Commission. (2010). Future Internet for future European economies and societies.

Fan, Z., Kulkarni, P., Gormus, S., Efthymiou, C., Kalogridis, G., Sooriyabandara, M., ... Chin, W. H. (2012). Smart grid communications: Overview of research challenges, solutions, and standardization activities. *IEEE Communications Surveys & Tutorials*, *15*(1), 21–38.

Fang, X., Misra, S., Xue, G., & Yang, D. (2011). Smart grid—The new and improved power grid: A survey. *IEEE Communications Surveys & Tutorials*, *14*(4), 944–980.

Finster, M., Phillips, J., & Wallace, K. (2016). Front-line resilience perspectives: The electric grid (No. ANL/GSS-16/2). Argonne, IL (United States): Argonne National Lab (ANL).

Flick, T., & Morehouse, J. (2010). *Securing the smart grid: Next generation power grid security*. Elsevier.

Franceschetti, G., & Grossi, M. (Eds.), (2011). *Homeland security threats, countermeasures, and privacy issues*. Artech House.

Giordano, V., & Fulli, G. (2012). A business case for smart grid technologies: A systemic perspective. *Energy Policy*, *40*, 252–259.

Grand, J. (2005). Protecting your crown jewels: An introduction to embedded security for hardware-based products. *Computer Fraud & Security*, *2005*(10), 13–20.

Gunduz, M. Z., & Das, R. (2020). Cyber-security on smart grid: Threats and potential solutions. *Computer Networks*, *169*, 107094.

Gungor, V. C., & Lambert, F. C. (2006). A survey on communication networks for electric system automation. *Computer Networks*, *50*(7), 877–897.

Kesan, J. P., & Hayes, C. M. (2016). Bugs in the market: Creating a legitimate, transparent, and vendor-focused market for software vulnerabilities. *Arizona Law Review*, *58*, 753.

Khan, A. A., Kumar, V., Ahmad, M., & Rana, S. (2021). LAKAF: Lightweight authentication and key agreement framework for smart grid network. *Journal of Systems Architecture*, *116*, 102053.

Kothari, D. P., & Nagrath, I. J. (2003). *Modern power system analysis*. Tata McGraw-Hill Education.

Kumari, A., & Tanwar, S. (2020). Secure data analytics for smart grid systems in a sustainable smart city: Challenges, solutions, and future directions. *Sustainable computing: Informatics and systems*, *28*, 100427.

Luhmann, N. (2018). *Trust and power*. John Wiley & Sons.

Möllering, G. (2001). The nature of trust: From Georg Simmel to a theory of expectation, interpretation and suspension. *Sociology*, *35*(2), 403–420.

Möllering, G. (2006). Trust beyond risk The leap of Faith. *Trust: Reason, routine, reflexivity*, 105–126.

Odelu, V., Das, A. K., Wazid, M., & Conti, M. (2016). Provably secure authenticated key agreement scheme for smart grid. *IEEE Transactions on Smart Grid*, *9*(3), 1900–1910.

Patel, A., Alhussian, H., Pedersen, J. M., Bounabat, B., Júnior, J. C., & Katsikas, S. (2017). A nifty collaborative intrusion detection and prevention architecture for smart grid ecosystems. *Computers & Security*, *64*, 92–109.

Pérez-Guzmán, R. E., Salgueiro-Sicilia, Y., & Rivera, M. (2017, October). Communications in smart grids. In *2017 CHILEAN Conference on Electrical, Electronics Engineering, Information and Communication Technologies (CHILECON)* (pp. 1–7). IEEE.

Qi, M., & Chen, J. (2020). Two-pass privacy preserving authenticated key agreement scheme for smart grid. *IEEE Systems Journal.*

Quinn, E. L. (2009). Smart metering and privacy: Existing laws and competing policies. Available at SSRN 1462285.

Rahimi, F., & Ipakchi, A. (2010). Demand response as a market resource under the smart grid paradigm. *IEEE Transactions on Smart Grid, 1*(1), 82–88.

Rivas, A. E. L., & Abrao, T. (2020). Faults in smart grid systems: Monitoring, detection and classification. *Electric Power Systems Research, 189*, 106602.

Rizwan Asghar, M., Ion, M., Russello, G., & Crispo, B. (2013). ESPOON: Enforcing Encrypted Security Policies in Outsourced Environments. arXiv e-prints, arXiv-1306.

Schwartz, P. M. (2010). Data protection law and the ethical use of analytics. The Center for Information Policy Leadership, Hunton and Williams LLP. Retrieved from: privacyassociation.org/media/pdf/knowledge_center/Ethical_Underpinnings_of_Analytics.pdf

Simmhan, Y., Kumbhare, A. G., Cao, B., & Prasanna, V. (2011, July). An analysis of security and privacy issues in smart grid software architectures on clouds. In *2011 IEEE 4th International Conference on Cloud Computing* (pp. 582–589). IEEE.

Sundstrom, O., Gantenbein, D., & Binding, C. (2011). Adding Gridwise information to Ecogrid deliverable D1. 1. *IBM, Tech. Rep.*

Tene, O., & Wolf, C. (2013, January). Overextended: Jurisdiction and applicable law under the EU general data protection regulation. In *Future of Privacy Forum White Paper.*

Xu, H., Yu, W., Griffith, D., & Golmie, N. (2018). A survey on industrial Internet of Things: A cyber-physical systems perspective. *IEEE Access, 6*, 78238–78259.

Yan, L., Chang, Y., & Zhang, S. (2017). A lightweight authentication and key agreement scheme for smart grid. *International Journal of Distributed Sensor Networks, 13*(2), 1550147717694173.

Yan, Y., Qian, Y., Sharif, H., & Tipper, D. (2012). A survey on cyber security for smart grid communications. *IEEE Communications Surveys & Tutorials, 14*(4), 998–1010.

Index

Note: Page numbers followed by "*f*" and "*t*" refer to figures and tables, respectively.

E

R

S

T

FSC
www.fsc.org
MIX
Papier | Fördert gute Waldnutzung
FSC® C083411

Kolibri 360 GmbH
Ferdinand-Jühlke-Straße 7
99095 Erfurt, Deutschland
produktsicherheit@kolibri360.de